I0470664

September 2012

BIOSURVEILLANCE

DHS Should Reevaluate Mission Need and Alternatives before Proceeding with BioWatch Generation-3 Acquisition

GAO
Accountability ★ Integrity ★ Reliability

BIOSURVEILLANCE

DHS Should Reevaluate Mission Need and Alternatives before Proceeding with BioWatch Generation-3 Acquisition

GAO
Accountability * Integrity * Reliability

Highlights

Highlights of GAO-12-810, a report to congressional requesters

Why GAO Did This Study

The 2001 anthrax attacks brought attention to the potentially devastating consequences of a biological attack. DHS operates a program, known as BioWatch, intended to help detect such an attack by airborne pathogens. The currently deployed technology can take 12 to 36 hours to confirm the presence of pathogens. DHS has been pursuing a third generation of the technology that will perform automated testing, potentially generating a result in under 6 hours and reducing labor costs.

GAO was asked to examine issues related to the Gen-3 acquisition. This report addresses the extent to which (1) DHS used its acquisition life cycle framework to justify the need and consider alternatives; (2) DHS developed reliable performance, schedule, and cost expectations; and (3) steps remaining before Gen-3 can be deployed. GAO reviewed acquisition documentation and test results and interviewed agency officials from the BioWatch program and other DHS components with development, policy, and acquisition responsibilities.

What GAO Recommends

GAO recommends that before continuing the acquisition, DHS reevaluate the mission need and alternatives and develop performance, schedule, and cost information in accordance with guidance and good acquisition practices. DHS concurred with the recommendations, but not the implementation timeline. DHS plans to proceed with the acquisition while implementing them to avoid further delays. However, GAO believes the recommendations should be enacted before DHS proceeds with the acquisition as discussed in this report.

View GAO-12-810. For more information, contact Bill Jenkins at (202) 512-8757 or jenkinswo@gao.gov.

What GAO Found

The Department of Homeland Security (DHS) approved the Generation-3 (Gen-3) acquisition in October 2009, but it did not fully engage in the early phases of its acquisition framework to ensure that the acquisition was grounded in a justified mission need and that it pursued an optimal solution. Critical processes in the early phases of DHS's framework are designed to (1) justify a mission need that warrants investment of resources and (2) select an optimal solution by evaluating viable alternatives based on risk, costs, and benefits. BioWatch program officials said that these early acquisition efforts were less comprehensive and systematic than the DHS framework calls for because there was already departmental consensus around the solution. Without a systematic effort to justify the need for the acquisition in the context of its costs, benefits, and risks, DHS has pursued goals and requirements for Gen-3 with limited assurance that they represent an optimal solution. Reevaluating the mission need and systematically analyzing alternatives could provide better assurance of an optimal solution.

The performance, schedule, and cost expectations presented in required documents when DHS approved the acquisition were not developed in accordance with DHS guidance and good acquisition practices—like accounting for risk in schedule and cost estimates. BioWatch program officials said that DHS leadership directed them to prepare information quickly for the 2009 decision, which was accelerated by more than 1 year. Since DHS approved the acquisition in October 2009, the estimated date for full deployment has been delayed from fiscal year 2016 to fiscal year 2022, and the original life cycle cost estimate for the 2009 decision—a point estimate unadjusted for risk—was $2.1 billion. In June 2011, DHS provided a risk-adjusted estimate at the 80 percent confidence level of $5.8 billion. Comprehensive and systematic information developed using good practices for cost and schedule estimating, could help ensure more reliable performance, schedule, and cost information for decision makers.

Several steps remain before DHS can deploy and operate Gen-3. First, DHS must conduct additional performance and operational testing. This testing—estimated to take 3 years and cost $89 million—is intended to demonstrate full system performance, including the information technology network. To do so, the BioWatch program must address testing challenges including limitations on the use of live pathogens, among others. Following operational testing, DHS intends to decide whether to authorize the production and deployment of Gen-3. If Gen-3 is approved, the BioWatch program plans to prepare for deployment by working with BioWatch jurisdictions to develop location-specific plans to guide Gen-3 operations. DHS estimates show that about $5.7 billion of the $5.8 billion life-cycle cost remains to be spent to test, produce, deploy, and operate Gen-3 through fiscal year 2028.

Previous Spending on Gen-3 and Estimated Costs Remaining

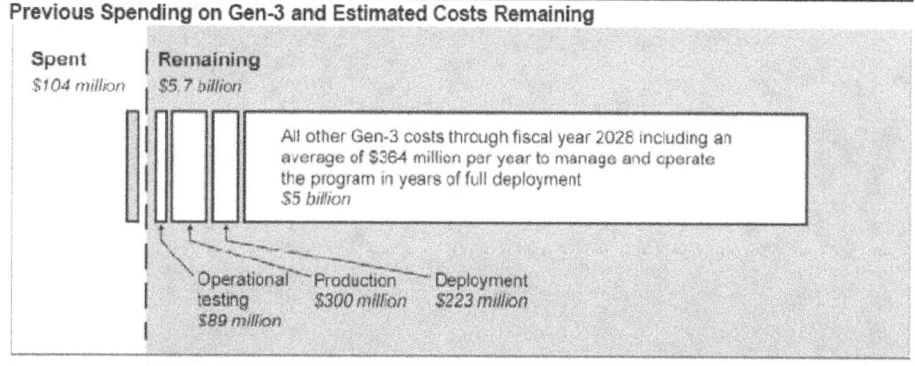

Source: GAO analysis of BioWatch program documents.

Contents

Figures

Abbreviations

ADE	Acquisition Decision Event
AMD	Acquisition Management Directive
APDS	Autonomous Pathogen Detection System
BAND	Bioagent Autonomous Network Detector
BAR	BioWatch Actionable Result
DHS	Department of Homeland Security
DOD	Department of Defense
DNA	deoxyribonucleic acid
GEN-2	BioWatch Generation-2
GEN-3	BioWatch Generation-3
HSPD-10	Homeland Security Presidential Directive-10
IPG	Integrated Planning Guidance
KPP	Key Performance Parameter
LOD	Limit of Detection
M-BAND	Biological Autonomous Network Detector (new name of BAND)
NG-ADS	Next Generation Automated Detection System
OHA	Office of Health Affairs
OMB	Office of Management and Budget
PARM	Program Accountability and Risk Management
PCR	polymerase chain reaction
S&T	Science and Technology Directorate
TRL	Technology Readiness Level

September 10, 2012

The Honorable Susan M. Collins
Ranking Member
Committee on Homeland Security
    and Governmental Affairs
United States Senate

The Honorable M. Gus Bilirakis
Chairman
Subcommittee on Emergency Preparedness, Response,
    and Communication
The Honorable Daniel E. Lungren
Chairman
Subcommittee on Cybersecurity, Infrastructure Protection,
    and Security Technologies
Committee on Homeland Security
House of Representatives

Soon after the terrorist attacks of September 11, 2001, letters laced with anthrax began appearing in the U.S. mail. During the incident—known as Amerithrax—5 Americans were killed and 17 were sickened in what became one of the worst biological attacks in U.S. history. The intentional use of a biological agent in this manner brought new awareness of the threat posed by and the potentially devastating consequences of a biological weapon. This awareness led to increased focus on developing the ability to provide early detection of and situational awareness during a disease outbreak. As we have reported in June 2010 and again in October 2011, federal efforts to combat the risk of a biological attack have included enhancing traditional public health surveillance and developing new approaches to prevention, early detection, and treatment. However, as we also noted, these efforts have not been guided by an overarching strategy and leadership to help ensure that they complement each other to efficiently and effectively mitigate as much risk as possible.[1] In 2011, the National Academies' Institute of Medicine and National Research Council cautioned that new approaches to biosurveillance need to be integrated into the overall prevention and response system, and their cost-effectiveness needs to be evaluated in comparison with other investments that could be made to attain the same goals.[2]

According to Homeland Security Presidential Directive 10 (HSPD-10): Biodefense for the 21st Century, a national bioawareness capability providing early warning, detection, or recognition of a biological weapon attack is an essential component of biodefense.[3] To contribute to this national capability, the Department of Homeland Security (DHS) operates the BioWatch program, which uses routine laboratory testing designed to detect an aerosolized biological attack for five specific biological agents considered high risk for use as biological weapons. The BioWatch program is a federally-managed, locally operated system with collectors deployed primarily in outdoor locations in more than 30 major metropolitan areas. The technology currently deployed relies on capturing

---

[1]GAO, *Biosurveillance: Efforts to Develop a National Biosurveillance Capability Need a National Strategy and a Designated Leader,* GAO-10-645 (Washington, D.C.: June 30, 2010), and *Biosurveillance: Nonfederal Capabilities Should Be Considered in Creating a National Biosurveillance Strategy,* GAO-12-55 (Washington, D.C.: Oct. 31, 2011). In June 2010, GAO recommended that the Homeland Security Council direct the National Security Staff within the White House create a focal point to lead development of a national strategy. In October 2011, we reiterated that recommendation and further recommended that such a strategy should include specific efforts to account for the challenges associated with federal reliance on state and local partnerships. On July 31, 2012, the White House released the *National Strategy for Biosurveillance* to describe the U.S. government's approach to strengthen the biosurveillance enterprise. The strategy describes guiding principles, core functions, and enablers for strengthening biosurveillance. A strategic implementation plan is to be completed within 120 days of the strategy issuance. According to the strategy, the implementation plan will include specific actions and activity scope, designated roles and responsibilities, and a mechanism for evaluating progress.

[2]Institute of Medicine and National Research Council of the National Academies, Committee on Effectiveness of National Biosurveillance Systems, *BioWatch and the Public Health System, BioWatch and Public Health Surveillance: Evaluating Systems for the Early Detection of Biological Threats* (Washington, D.C.: 2011). The *National Strategy for Biosurveillance* defines "biosurveillance" as the process of gathering, integrating, interpreting, and communicating essential information related to all-hazards threats or disease activity affecting human, animal, or plant health to achieve early detection and warning, contribute to overall situational awareness of the health aspects of an incident, and enable better decision making at all levels.

[3]HSPD-10*: Biodefense for the 21st Century* (Washington, D.C., April 2004).

air samples, which are then manually collected and transported to a laboratory for testing—a process that can take 12-36 hours and entails labor costs for manual collection and analysis. DHS has been pursuing a new technology—which is to be the third generation of deployed BioWatch technology, called BioWatch Generation-3 (Gen-3).[4] The goal of Gen-3 is to improve upon existing technology by enabling autonomous collection and analysis of air samples using the same laboratory science that is carried out in manual processes to operate the current system. In essence, the new technology is to operate as a self-contained "laboratory-in-a-box" that, compared with the current system, would reduce the time between potential exposure and confirmation of the presence of biological pathogens and eliminate manual collection and analysis costs.

BioWatch Gen-3 has a history of technical and management challenges. With a current estimated life-cycle cost of $5.8 billion (risk adjusted at the 80 percent confidence level) to procure and operate, it is one of DHS's

---

[4]Throughout this report the terminology "Gen-3" is used to refer to the set of DHS efforts to develop, acquire, and deploy an autonomous, networked aerosolized biodetection technology. Efforts to acquire and deploy Gen-3 technology began under DHS's Office of Health Affairs in the late 2007 to early 2008 time frame. However, efforts to develop networked aerosolized biodetectors to be used by the BioWatch program began as early as 2003, when DHS's Science and Technology Directorate issued a research announcement for its Bioagent Autonomous Network Detector (BAND) and related efforts. From 2007 through 2009, DHS was involved with two separate technological approaches to autonomous detection. One—referred to at that time as Gen-3—was the technology that resulted from the BAND effort. The other—called Gen-2.5, or Autonomous Pathogen Detection System (APDS)—was a deployed prototype developed by Lawrence Livermore National Laboratory—in association with private industry—that DHS had considered deploying as an interim solution until the Gen-3 acquisition was completed. In October 2008, after determining that before it could be fully deployed as an interim solution, APDS would need nearly as much time in testing as the BAND-based technology, DHS decided that both technologies should be tested simultaneously. In the same period, language in the committee print accompanying the DHS appropriation act called for a competitive bid process for the first testing phase of the Gen-3 acquisition. Staff of House Comm. On Appropriations, 110th Cong. 656 (Comm. Print 2009). Five vendors responded to the request for proposal and DHS awarded contracts to two—the vendor that manufactured BAND (now known as M-BAND) and the vendor that used the APDS-based system (now known as the Next Generation Automated Detection System, or NG-ADS). At this point, both became Gen-3 candidate technologies, but only NG-ADS completed the first round of testing.

largest ongoing acquisitions.[5] DHS will soon be confronted with deciding whether or not to implement the next stage of the acquisition, which includes initial production, testing, full production, and deployment. We have previously reported on acquisition challenges at DHS, and though DHS has taken steps to improve its acquisition management, it remains on our High-Risk List as part of the department's efforts to strengthen and integrate its management functions.[6]

DHS acquisition efforts are guided by a four-phased template for planning and executing acquisitions. Among other purposes, this template serves to help DHS ensure that acquisitions respond to a justified mission need; to select the optimal alternative while balancing cost, schedule, and risk realities; and to ensure that reliable performance, schedule, and cost estimates are available to help guide decision making in every phase. Expressing questions about whether DHS had undertaken rigorous efforts to help guide such decision making, you asked us to examine several issues related to the development, acquisition, and deployment of Gen-3. This report addresses the following questions:

---

[5]Estimated Gen-3 costs in this report are based on DHS's June 2011 Life-Cycle Cost Estimate, which include estimated costs through fiscal year 2028. These costs were risk adjusted to the 80 percent confidence level. The cost estimate also included a second risk-adjusted estimate at the 50 percent confidence level of $4.5 billion and a point estimate that reflects the 28 percent confidence level of $3.8 billion.

[6]Every 2 years, we provide Congress with an update on our High-Risk Program, which highlights major problems that are at high risk for waste, fraud, abuse, mismanagement or in need of broad reform. DHS acquisition is one of several management functions—along with the integration of those functions—that fall under the Implementing and Transforming DHS category in the High-Risk Series. In the most recent update in 2011, we noted that DHS revised its acquisition management oversight policies to include detailed guidance to inform departmental acquisition decision making. However, DHS's Acquisition Review Board had not reviewed most major programs, and DHS did not yet have accurate cost estimates for most programs. See GAO, *High-Risk Series: An Update*, GAO-11-278 (Washington, D.C.: Feb. 16, 2011). In March 2012, we testified that DHS had made progress addressing management challenges in some key areas, including some acquisition processes and practices; however, we also noted that DHS had considerable work ahead to implement corrective action plans and address management challenges. See GAO, *Department of Homeland Security: Continued Progress Made Improving and Integrating Management Areas, but More Work Remains*, GAO-12-365T (Washington, D.C.: Mar. 1, 2012).

1. In making its initial decision to acquire Gen-3, to what extent did DHS use its Acquisition Life-cycle Framework processes to justify the need and consider alternatives for Gen-3?

2. To what extent did DHS ground the Gen-3 acquisition in reliable performance, schedule, and cost expectations?

3. What remains to be done before Gen-3 technology can be produced and deployed?

To address our objectives and to determine the specific requirements and the broader intent of the department's acquisition management processes, we reviewed DHS's Acquisition Life-cycle Framework—the department's four-phased template for planning and executing acquisitions that is described in DHS's Acquisition Management Directive (AMD) 102-01 and associated instructional guidebook.[7] We focused on activities related to the acquisition of Gen-3 since fiscal year 2007—when DHS's Office of Health Affairs (OHA) began to manage the BioWatch program. To determine the extent to which DHS used the Acquisition Life-cycle Framework processes to help ensure the decision to invest in Gen-3 was informed by analyses of alternatives; costs and benefits; and reliable performance, schedule, and cost expectations, we reviewed DHS's Acquisition Life-cycle Framework guidance for the first two acquisition phases—identifying a capability need and analyzing and selecting the means to provide that capability. This included guidance for engaging in acquisition processes and preparing related acquisition documents such as the Mission Needs Statements, Analyses of Alternatives, Operational Requirement Documents, Life-Cycle Cost Estimates, and Acquisition Program Baselines. We compared documentation prepared for the Gen-3 acquisition with the guidance. Where appropriate, we also consulted external guidance to assess the extent to which key acquisition documents were produced in a reliable

---

[7]DHS first issued the AMD-102-01 in November 2008. Prior to November 2008, DHS operated under the March 2006 Management Directive No. 1400 on the Investment Review Process. Because some acquisition activities that we evaluated could have or should have commenced prior to the issuance of AMD-102-01, we consulted Directive 1400 to ensure that the basic principles of acquisition management we used to evaluate DHS's actions against—particularly for events before or near November 2008—were consistent in both sets of guidance.

manner, including the *GAO Cost Estimating and Assessment Guide*.[8] To provide the appropriate historical context to our assessment of the 2009 documentation set for the Gen-3 acquisition, we consulted with DHS officials in a variety of offices, including the OHA BioWatch Program Office—the group sponsoring the Gen-3 acquisition—the Science and Technology Directorate, the DHS Office of Policy, and the Program Accountability and Risk Management (PARM) office, which manages the oversight of acquisition programs and implements DHS acquisitions guidance and reports directly to the DHS Undersecretary for Management.

To understand the challenges the program faced in preparing the 2009 documentation set, we consulted with BioWatch and PARM officials and reviewed external reviews of the program, including a 2011 National Academies of Science Report and a 2012 report from Sandia National Laboratories.[9] We did not complete a thorough assessment and validation of these studies; however, we reviewed the studies and found them appropriate for the purposes used in this report. We also compared the 2009 Gen-3 documentation set with the most recent versions of these documents to assess the extent to which performance, schedule, and cost expectations have changed.

To determine what remains to be done before the Gen-3 technology can be produced and deployed, we compared the current status of the acquisition with DHS's Acquisition Life-cycle Framework requirements. To determine the current status of the acquisition, we reviewed program documentation since fiscal year 2008—including Acquisition Decision Memorandums that summarize program reviews conducted by DHS— and consulted with BioWatch program and PARM officials. To understand what additional steps beyond the Acquisition Life-cycle Framework requirements must be completed, as well as any challenges the program faces in completing the acquisition and deploying the technology, we consulted with BioWatch and PARM officials, reviewed test results from the initial round of testing completed in 2010 and 2011, and reviewed

---

[8]GAO, *GAO Cost Estimating and Assessment Guide*, GAO-09-3SP (Washington, D.C.: Mar. 2, 2009).

[9]See Institute of Medicine and National Research Council, *BioWatch and Public Health Surveillance*, 2011, and *BioWatch Technical Analysis of Biodetection Architecture Performance*, Sandia National Laboratories, January 2012.

plans for future testing. Finally, to quantify the level of resources required and the time that remains to complete the full deployment of Gen-3, we reviewed program documentation including the Life-Cycle Cost Estimate and Acquisitions Program Baseline.

We conducted this performance audit from January 2012 to September 2012 in accordance with generally accepted government auditing standards. Those standards require that we plan and perform the audit to obtain sufficient, appropriate evidence to provide a reasonable basis for our findings and conclusions based on our audit objectives. We believe that the evidence obtained provides a reasonable basis for our findings and conclusions based on our audit objectives.

Background

BioWatch and the Biodefense Enterprise

Biological threats that could result in catastrophic consequences exist in many forms and arise from multiple sources. For example, several known biological agents could be made into aerosolized weapons and intentionally released in a transportation hub or other populated urban setting, introduced into the agricultural infrastructure and food supply, or used to contaminate the water supply. Concerned with the threat of bioterrorism, in 2004, the White House released HSPD-10, which outlines four pillars of the biodefense enterprise and discusses various federal efforts and responsibilities that help to support it. The biodefense enterprise is the whole combination of systems at every level of government and the private sector that can contribute to protecting the nation and its citizens from potentially catastrophic effects of a biological event. It is composed of a complex collection of federal, state, local, tribal, territorial, and private resources, programs, and initiatives, designed for different purposes and dedicated to mitigating various risks, both natural and intentional.

The four pillars of biodefense outlined in HSPD-10 are (1) threat awareness, (2) prevention and protection, (3) surveillance and detection, and (4) response and recovery. Protecting humans, animals, plants, air, soil, water, and critical infrastructure from potentially catastrophic effects of intentional or natural biological events entails numerous activities carried out within and between multiple federal agencies and their nonfederal partners. Figure 1 shows the four pillars of biodefense, examples of some federal efforts that can support them, and federal agencies responsible for those efforts. The BioWatch program falls under the surveillance and detection pillar. It is an example of an environmental monitoring activity.

Figure 1: The Pillars of Biodefense and Examples of Federal Departments and Efforts That Can Support Them

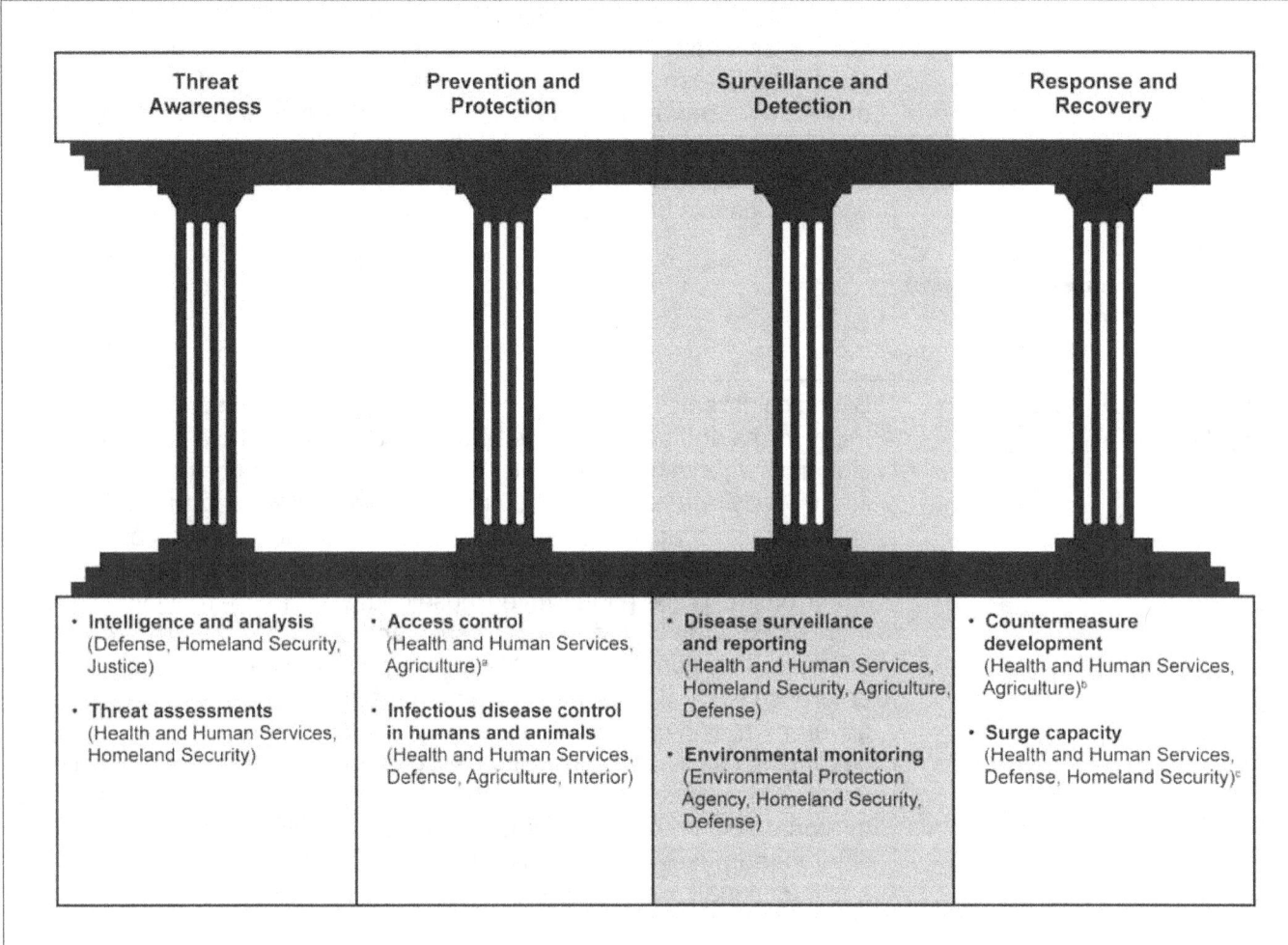

Source: GAO analysis of HSPD-10.

[a]Access control involves controlling access and use of pathogens by those who would use them to do harm.

[b]Countermeasure development involves the development and availability of sufficient quantities of safe and efficacious medical countermeasures to mitigate illness and death.

[c]Surge capacity involves ensuring that existing medical infrastructure such as hospitals, laboratories, and emergency medical services have capacity to respond to major disasters, such as a potentially catastrophic biological event.

BioWatch Past and Present

DHS, in cooperation with other federal agencies, created the BioWatch program in 2003.[10] The goal of BioWatch is to provide early warning, detection, or recognition of a biological attack. When DHS was established in 2002, a perceived urgency to deploy useful—even if imperfect—technologies in the face of potentially catastrophic consequences catalyzed the rapid deployment of many technologies, including the earlier generations of BioWatch collectors. In the initial deployment of BioWatch collectors—known as Generation 1—DHS deployed detectors to 20 major metropolitan areas, known as BioWatch jurisdictions, to monitor primarily outdoor spaces.[11] DHS completed this initial deployment quickly—within 80 days of the President's announcement of the BioWatch program during his 2003 State of the Union Address. To accomplish this quick deployment, DHS adapted an existing technology that was already used to accomplish other air monitoring missions. In 2005, DHS expanded BioWatch to an additional 10 jurisdictions, for a total of 30. This expanded deployment—referred to as Generation 2 (Gen-2)—also included the addition of indoor monitoring capabilities in three high-threat jurisdictions and provided additional capacity for events of national significance, such as major sporting events and political conventions.

Currently, the BioWatch program collaborates with 30 BioWatch jurisdictions throughout the nation to operate approximately 600 Gen-2 collectors. These detectors rely on a vacuum-based collection system that draws air samples through a filter. These filters must be manually collected and transported to state and local public health laboratories for analysis using a process called Polymerase Chain Reaction (PCR).[12]

---

[10]The BioWatch program was established on January 10, 2003, and is currently managed by DHS's Office of Health Affairs. Prior to 2007, the BioWatch program was managed by DHS's Science and Technology Directorate.

[11]Each BioWatch jurisdiction may include various state and local government entities, such as counties or cities, or support contractors.

[12]Sometimes called molecular photocopying, the PCR is a fast and inexpensive technique used to amplify (or copy) small segments of deoxyribonucleic acid (DNA) to support analyses such as detecting the DNA sequences in the five agents the BioWatch program is designed to detect. To amplify a segment of DNA, the sample is heated so the DNA separates into two pieces of single-stranded DNA. Then, an enzyme builds two new strands of DNA, using the original strands as templates. This process results in the duplication of the original DNA, containing one old and one new strand of DNA. Each of these strands can be used to create two more copies. The cycle can be repeated as many as 30 or 40 times, until enough genetic material is available for analysis.

During this process, the sample is evaluated for the presence of genetic material from five different biological agents. If genetic material is detected, a BioWatch Actionable Result (BAR) is declared.[13] Using this manual process, the determination of a BAR can occur from 12 to 36 hours after an agent is initially captured by the air filter. This 36-hour timeline consists of up to 24 hours for air sampling, up to 4 hours for sample recovery, and up to 8 hours for laboratory testing.

BioWatch in Action

Each BioWatch jurisdiction has either a BioWatch Advisory Committee or equivalent decision making group in place, composed of public health officials, first responders, and other relevant stakeholders. The BioWatch Advisory Committee is responsible for the day-to-day BioWatch operations, including routine filter collection and laboratory analysis of filter samples. In the event of a BAR, the BioWatch Advisory Committee is also responsible for determining whether that BAR poses a public health risk and deciding how to respond. The declaration of a BAR does not necessarily signal that a biological attack has occurred, as the Gen-2 detection process is highly sensitive and can detect minute amounts of pathogens that naturally occur in the environment. For example, at least two of the agents the program monitors occur naturally and have been detected in numerous areas of the United States. Since 2003, more than 100 BARs have been declared according to BioWatch program officials, but none were determined to be a potential risk to public health. Figure 2 shows the process that local BioWatch jurisdictions are to follow when deciding how to respond to a BAR.

---

[13]The BioWatch program defines a BAR as one or more PCR-verified positive results from a single BioWatch collector. A positive result requires multiple strands of the PCR-amplified DNA to match an algorithm that has been designed to indicate the presence of genetic material from one or more of the five agents in question.

Figure 2: Process Used By Jurisdictions to Detect and Respond to a BioWatch Actionable Result

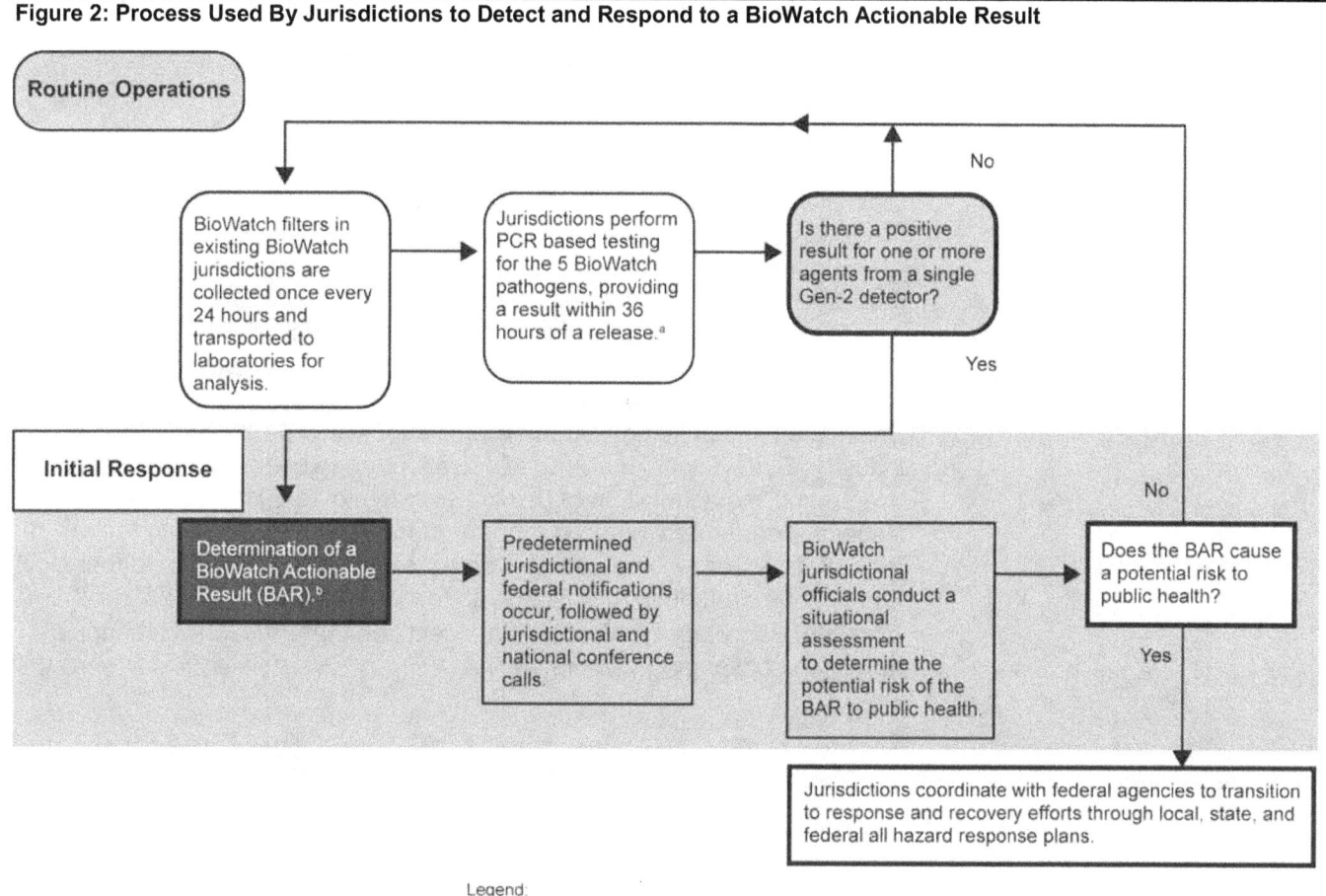

Legend:
PCR - Polymerase Chain Reaction

Source: GAO analysis of BioWatch program guidance.

[a]PCR is a technique to copy DNA for laboratory testing.

[b]The BioWatch program defines a BAR as one or more PCR-verified positive results from a single BioWatch collector. A positive result requires multiple strands of the PCR-amplified DNA to match an algorithm that has been designed to indicate the presence of genetic material from one or more of the five agents in question.

Gen-3 Development and Challenges

To reduce the time required to detect biological pathogens, DHS has been pursuing an autonomous detection capability for the BioWatch program. Envisioned as a laboratory-in-a-box, the autonomous detection system that DHS seeks would automatically collect air samples, produce and read PCR results every 4 to 6 hours, and communicate the results to public health officials without manual intervention. By automating the analysis, DHS anticipates that detection time could be reduced to 6 hours or less, making the technology more appropriate for monitoring indoor high-throughput facilities such as transportation nodes. DHS also anticipates that operational costs will be reduced through the elimination of the daily manual collection and laboratory analysis process.

Developing autonomous detection has proved challenging according to BioWatch program officials, in part because some of the technology required was novel, but also because even the existing technologies—for example, the air collection system and the apparatus that reads the PCR results—had not been combined for this specific application in an operational environment before. As shown in figure 3, DHS began to develop autonomous detection technology in 2003. Initially, development of technologies to support autonomous detection was led by DHS's Science and Technology Directorate (S&T), which partnered with industry. Since fiscal year 2007, DHS's OHA has been responsible for overseeing the acquisition of this technology. In its 2011 report, the National Academies reported that the proposed enhancements to the BioWatch system will be possible only if significant scientific and technical hurdles are overcome.[14] Similarly, as recently as March 2012, DHS's Assistant Secretary for Health Affairs testified that the Gen-3 technology has been challenging to develop.[15]

---

[14]See Institute of Medicine and National Research Council, *BioWatch and Public Health Surveillance,* 2011.

[15]House Homeland Security Committee, Hearing on 2013 Budget: DHS Office of Health Affairs (Washington D.C.: Mar. 29, 2012).

Figure 3: Timeline of Investments and Activities for Establishment of BioWatch and Early Gen-3 Activities

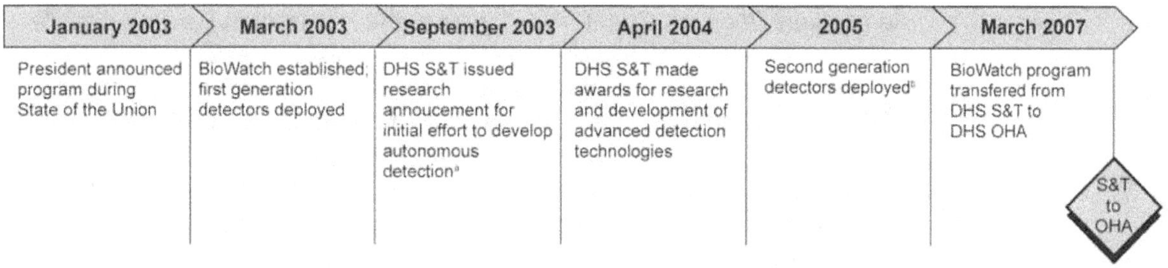

January 2003	March 2003	September 2003	April 2004	2005	March 2007
President announced program during State of the Union	BioWatch established; first generation detectors deployed	DHS S&T issued research annoucement for initial effort to develop autonomous detection[a]	DHS S&T made awards for research and development of advanced detection technologies	Second generation detectors deployed[b]	BioWatch program transfered from DHS S&T to DHS OHA

S&T to OHA

Legend:
S&T - Department of Homeland Security Science and Technology Directorate
OHA - Department of Homeland Security Office of Health Affairs
Source: GAO analysis of historic BioWatch program information.

[a]DHS Research Announcement 03-01 (September 23, 2003).

[b]According to BioWatch program officials there was no significant technological difference between first-generation and second-generation BioWatch detectors; the primary difference was the larger area covered by all the deployed detectors.

DHS's Four-Phase Acquisition Life-cycle Framework

The overall policy and structure for acquisition management outlined in DHS's AMD 102-01 includes the department's Acquisition Life-cycle Framework—a template for planning and executing acquisitions. According to the directive, DHS adopted the Acquisition Life-cycle Framework to ensure consistent and efficient acquisition management, support, review, and approval throughout the department. As we have previously reported, without the development, review, and approval of key acquisition documents, agencies are at risk of having poorly defined requirements that can negatively affect program performance and contribute to increased costs.[16]

As shown in figure 4, DHS's Acquisition Life-cycle Framework includes four acquisition phases through which DHS determines whether it is sensible to proceed with a proposed acquisition: (1) identify a capability need; (2) analyze and select the optimal solution to meet that need; (3) obtain the solution; and (4) produce, deploy, and support the solution. During the first three phases, the DHS component pursuing the

---

[16]GAO, *Homeland Security: DHS Could Strengthen Acquisitions and Development of New Technology*, GAO-11-829T (Washington, D.C.: July 15, 2011).

acquisition is required to produce key documents in order to justify, plan, and execute the acquisition. These phases each culminate in an Acquisition Decision Event (ADE), where the Acquisition Review Board—a cross-component board of senior DHS officials—determines whether a proposed acquisition has met the requirements of the relevant Acquisition Life-cycle Framework phase and is able to proceed.[17] The Acquisition Review Board is chaired by the Acquisition Decision Authority—the official responsible for ensuring compliance with AMD 102-01. For the Gen-3 acquisition, DHS's Deputy Secretary serves as the Acquisition Decision Authority.

Figure 4: DHS's Acquisition Lifecycle Framework

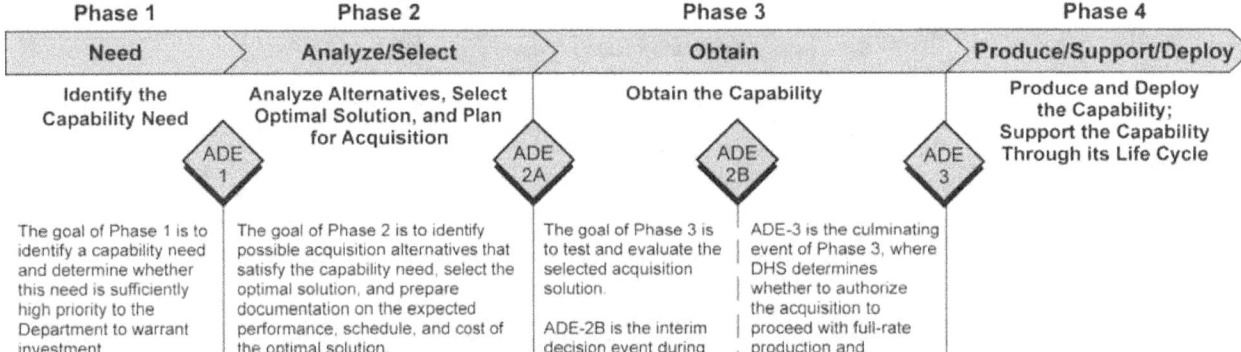

DHS held an Acquisition Review Board related to ADE-2B on August 16, 2012, during which the BioWatch program was seeking approval to

---

[17]Components represented on the Acquisition Review Board include the Office of Policy, the Science and Technology Directorate, General Counsel, and the Procurement Office, among others.

initiate the next phase of the acquisition.[18] The acquisition decision authority did not make a final ADE-2B decision, but did authorize the program to issue a solicitation for performance testing under the next testing phase. The Acquisition Decision Authority also required that the program office return to the Acquisition Review Board for approval prior to issuing a performance testing contract—which would allow the program to acquire a small number of test units. Furthermore, before undertaking the remaining steps in the acquisition, the program office must return to the Acquisition Review Board for ADE-2B with updated information, including an Analysis of Alternatives and Concept of Operations. DHS has not specified a time frame for completing these actions, but according to DHS officials, completing the Analysis of Alternatives may take up to 1 year.

DHS Did Not Fully Develop Critical Information for Initial Investment Decision Making

DHS approved the Gen-3 acquisition in October 2009, but it did not fully engage in the early phases of its acquisition framework to ensure that the acquisition was grounded in a justified mission need and that it pursued an optimal solution—for example, DHS did not fully develop a Mission Needs Statement or an Analysis of Alternatives with a cost-benefit analysis, as called for in its Acquisition Life-cycle Framework.

DHS Did Not Use Acquisition Life-cycle Framework Processes to Identify a Justified Mission Need

DHS skipped Phase 1 of the Acquisition Life-cycle Framework for the Gen-3 acquisition.[19] Specifically, it did not hold ADE-1 and prepared a Mission Needs Statement later to justify a predetermined solution. According to DHS's acquisition lifecycle framework, in Phase 1, the program office is to develop a Mission Needs Statement to make a case to decision makers that the acquisition represents a justified need that

[18]According to DHS officials, the remaining steps in the Gen-3 acquisition include performance testing, operational testing and evaluation, production, deployment, and sustainment.

[19]According to DHS officials, the Gen-3 acquisition was ongoing when Acquisition Management Directive 102-01 was issued. The officials said that many DHS programs that were ongoing in 2009 faced similar challenges. Nevertheless, DHS Management Directive 1400, which preceded Acquisition Management Directive 102-01, was similarly designed to, among other things, ensure that investments directly support and further DHS's missions. Management Directive 1400 also describes a phased lifecycle investment construct in which the first step is defining the mission need in a Mission Need Statement, which like the Mission Need Statement called for in Acquisition Management Directive 102-01 was to be a high-level description of a capability gap rather than a specific solution.

warrants the allocation of limited resources. At the end of Phase 1, the Acquisition Decision Authority is to review the Mission Needs Statement and other information during ADE-1 and decide whether the need is of sufficiently high priority to continue with the acquisition. However, according to BioWatch program officials, the Gen-3 acquisition began at ADE-2A, which is intended to be the decision gate at the end of Phase 2 of the Acquisition Lifecycle Framework. The Mission Needs Statement was finalized on October 6, 2009, just weeks before ADE-2A.

As shown in figure 5, DHS began to pursue a specific autonomous detection solution well before completing a Mission Needs Statement. Specifically, DHS's Integrated Planning Guidance (IPG) for fiscal year 2010-2014, which was finalized in March 2008, included very specific goals for the next generation of BioWatch—to deploy in all major cities an autonomous BioWatch detection device reducing the operating cost per site by more than 50 percent and warning time to less than 6 hours. The purpose of DHS's IPG is to communicate the Secretary's policy and planning goals to component-level decision makers to inform their programming, budgeting, and execution activities. As such, this specific set of goals for BioWatch Gen-3 demonstrates that DHS leadership had established a course for the acquisition by March 2008, in advance of any efforts to define the mission need through the Mission Needs Statement process, which was finalized more than a year and a half later.

BioWatch program officials said they were directed by DHS to prepare the Mission Needs Statement—along with other required documentation for the ADE-1 and ADE-2A decision gates—on an accelerated time frame of about 6 weeks to prepare for the ADE-2A decision, efforts that they said would typically require at least 8 months. According to these officials, they were aware that the Mission Needs Statement prepared for ADE-2A did not reflect a systematic effort to justify a capability need. Although such an effort would provide a platform to help make trade-off decisions in terms of costs, risks, and benefits throughout the remainder of the acquisition process, officials said the time they were given would not have allowed for such an effort. They said that the department directed them to proceed because there was already departmental consensus around the solution.

Moreover, in its fiscal year 2009 budget request, submitted in February 2008, DHS requested funding to procure BioWatch automated detection sensors and initiate deployment activities of the automated sensor system.[20] These funds, requested more than 18 months prior to the acquisition's formal approval at ADE-2A—were intended to fund operational testing activities for Gen-3 BioWatch prototypes as well as the procurement of 150 automated detection sensors that DHS sought to deploy as an interim solution until the full Gen-3 acquisition could be completed. A prototype version of this interim solution was first fielded in 2007, shortly after DHS's OHA assumed responsibility for the program.

Figure 5: Timeline of Activities and Investments for Acquisition Lifecycle Framework Phases 1 and 2

March 2007	December 2007	February 2008	March 2008	September 2008	September/October 2009
BioWatch program transfered to DHS OHA	DHS fielded prototypes as an interim autonomous detection solution	DHS requested first Gen-3 specific appropriation	DHS issued Integrated Planning Guidance containing specific goals for Gen-3	Appropriation conference report directed funds for Gen-3	Mission Needs Statement, Analysis of Alternatives, and other documentation prepared and submitted for ADE-2A decision gate

S&T to OHA

ADE 2A

Legend:
ADE - Acquisition Decision Event
S&T - Department of Homeland Security Science and Technology Directorate
OHA - Department of Homeland Security Office of Health Affairs

Source: GAO analysis of BioWatch program documents, DHS appropriations, and interviews with DHS officials.

Limited documentation is available to reflect the decision making process that occurred before the October 2009 Mission Needs Statement was finalized, including decisions related to the very specific IPG goals, the pursuit of funding for Gen-3, and the deployment of an interim solution before undertaking an effort to establish a justified mission need. We interviewed multiple officials in various DHS offices who had knowledge of Gen-3 in this early decision making period and the process that DHS used to justify the need to acquire Gen-3. However, none of these officials could describe what processes, if any, the department followed to determine that Gen-3 was a justified need. On the other hand, these officials all described a climate, in the wake of the September 11, 2001, terrorist attacks and the subsequent Amerithrax attacks, in which the highest levels of the administration expressed interest in quickly

---

[20]The conference report accompanying the fiscal year 2009 DHS appropriations act subsequently provided that $34.5 million of the appropriation was for BioWatch Gen-3. Staff of the House Comm. On Appropriations, 110th Cong. 656 (Comm. Print 2009).

deploying the early generation BioWatch detectors and subsequently improving the functionality of these detectors—as quickly as possible—to allow for faster detection and an indoor capability. On the basis of this interest, officials from the multiple DHS offices said it was their understanding that the administration and departmental leadership had already determined that the existing BioWatch technology would need to be expanded and entirely replaced with an autonomous solution well before the acquisition was approved at ADE-2A.

DHS guidance states that the Mission Needs Statement should consider the IPG, but it also directs the program to focus on the capability need without specifying a specific technical solution. The Mission Needs Statement is designed to serve as the foundational document upon which subsequent Acquisition Life-cycle Framework efforts are based. As such, a Mission Needs Statement that focuses on the capability need can help articulate and build consensus around the goals and objectives for a program in a way that provides a touchstone throughout the rest of the acquisition processes as the program endeavors to identify optimal solutions and contends with technology, budget, schedule, and risk realities. The Gen-3 Mission Needs Statement prepared for ADE-2A, in response to the very specific solution set prescribed by DHS leadership in the IPG, asserted a specific technological solution—total replacement of the existing vacuum-based, manual technology with autonomous detectors—as the only viable solution. Because the Mission Needs Statement was completed after DHS had prescribed specific goals for Gen-3 in the IPG and requested funding to field an interim solution, it appears to be justification for a predetermined solution, rather than the deliberate and systematic consideration of capability needs that would serve as the foundation for the remaining acquisition processes. As such, its utility as a foundation for subsequent acquisition efforts—for example identifying an optimal solution and balancing mission requirements with budget, schedule, and risk considerations—was limited.

DHS Did Not Use Acquisition Life-cycle Processes to Analyze Alternatives Based on Cost, Schedule, and Risk to Identify an Optimal Solution

DHS did not use the processes in Phase 2 of the Acquisition Life-cycle Framework to systematically identify the optimal solution based on cost-benefit and risk information. We have long advised that DHS make risk-informed investments of its limited resources. For example, in February 2005, we reported that because the nation cannot afford to protect everything against all threats, choices must be made about protection priorities given the risk and how to best allocate available resources.[21] More recently, we reported in September 2011 that because DHS does not have unlimited resources and cannot protect the nation from every conceivable threat, it must make risk-informed decisions regarding its homeland security approaches and strategies.[22] Phase 2 of the DHS Acquisition Life-cycle Framework is intended to support these kinds of trade-off decisions by requiring DHS components to complete an Analysis of Alternatives that systematically identifies possible alternative solutions that could satisfy the identified need, considers cost-benefit and risk information for each alternative, and finally selects the best option from among the alternatives.

The Analysis of Alternatives is intended to provide assurance to DHS at the ADE-2A decision gate that the component has chosen the most cost-effective solution to mitigate the capability gap identified in the Mission Needs Statement. To provide this assurance and allow DHS to make trade-off decisions, the guidance states that developing the Analysis of Alternatives should be a systematic analytic and decision making process to identify and document an optimal solution that includes an understanding of the costs and benefits of at least three viable alternatives.[23] The guidance directs the program to compare the alternatives based on cost, risk, and ability to respond to identified capability gaps. Finally, the guidance calls for an independent entity to complete this analysis to ensure that it is done objectively and without bias or vested interest in the study's outcome.

---

[21]GAO. *21st Century Challenges: Reexamining the Base of the Federal Government,*: GAO-05-325SP (Washington, D.C.: Feb. 1, 2005).

[22]GAO. *Department of Homeland Security: Progress Made and Work Remaining in Implementing Homeland Security Missions 10 Years after 9/11*, GAO-11-940T (Washington D.C.: Sept. 8, 2011).

[23]A viable alternative is one that responds to one or more capability needs as identified in the Mission Needs Statement.

The Gen-3 Analysis of Alternatives, completed in conjunction with the Mission Needs Statement by the BioWatch Program Office, does not reflect a systematic analytic and decision making process. Instead, the Analysis of Alternatives, like the Mission Needs Statement, was designed to support the decision the department had already made to pursue autonomous detection. The Analysis of Alternatives maintained that no modifications to the existing system would satisfy the goals in the IPG, and as such, it concluded that replacing the deployed technology entirely with autonomous detectors was the only viable solution. Along these lines, the Analysis of Alternatives included two alternatives:

1. expanding program coverage within current BioWatch cities and additional cities by replacing all currently deployed detectors with autonomous detection technology, and;

2. undertaking the same expansion using the currently deployed detectors but modifying the procedures to allow for filter collections every 8 hours instead of every 24 hours (a time frame that by definition would not meet the specific goals of the IPG).

The analysis did provide some cost information for each alternative, but it did not fully explore costs or consider benefit and risk information. As with the Mission Needs Statement, program officials told us that they were advised that a comprehensive Analysis of Alternatives would not be necessary because departmental consensus already existed that autonomous detection was the optimal solution.

In discussing the cost trade-offs between the autonomous detection solution and more frequent filter collection using the currently deployed technology in the Analysis of Alternatives, DHS focused on cost per detection cycle—that is, the cost each time an autonomous detector tests the air for pathogens or the cost each time a Gen-2 filter is manually collected and tested in a laboratory. According to our analysis of the June 2011 Life-Cycle Cost Estimate, cost per detection cycle is estimated to be seven times lower with the Gen-3 technology than the cost per detection cycle based on historical data for Gen-2 detectors. However, by only considering the cost per detection cycle of the two alternatives, the analysis does not help ensure the pursuit of an optimal solution based on cost, risk, and capability—as called for in the guidance.

To help ensure the optimal solution, DHS could benefit from a more complex and nuanced cost analysis that considers a number of factors in addition to the cost per detection cycle. For example, although not a cost-effectiveness analysis, table 1 shows that deploying and operating Gen-3 detectors is not necessarily more affordable than the existing Gen-2 deployment.[24] According to our analysis, the total annual cost to operate Gen-3 is estimated to be about four times more than the cost of the existing Gen-2 deployment. The higher cost reflects both higher annual operating costs per detector and an increase in the number of detectors and jurisdictions covered.

Table 1: Comparison of Gen-2 and Gen-3 Costs, Detectors, and Coverage (in 2010 dollars)

	Generation 2[a]	Generation 3
Annual operating costs		
Total program	$86.6 million	$363.6 million[b]
Number of detectors		
Number of collectors deployed and detectors planned for deployment	594	2,322
Annual cost per detector	$146,000	$157,000
Daily detection cycles		
Daily detection cycles per detector	1	8
Cost per detection cycle	$400	$54
Coverage		
Number of BioWatch jurisdictions	30	50
Total U.S. population covered	21 percent	33 percent
Population coverage in BioWatch jurisdictions	65 percent	90 percent

Source: GAO analysis of BioWatch program data.

[a]Gen-2 costs are total program costs based on an average of the fiscal year 2011 and fiscal year 2012 appropriation. According to BioWatch program officials, the 2-year appropriation average is the most accurate method for presenting Gen-2 costs.

[b]Gen-3 costs are total program costs based on an average of projected years of full deployment in the June 2011 Life-Cycle Cost Estimate. The June 2011 Life-Cycle Cost Estimate was presented in 2010 dollars, so the Gen-2 costs have been adjusted to correspond. Costs for Gen-3 generally do not include costs to operate Gen-2 during the transition to Gen-3. According to program officials, the two systems will run concurrently for up to 6 months in each of the jurisdictions in which Gen-3 is deployed.

---

[24]According to the Office of Management and Budget (OMB) Circular A-94, Appendix A, a cost-effectiveness analysis is a systematic method for comparing the costs of alternative means of achieving the same stream of benefits or a given objective.

In addition to a limited cost analysis, the Gen-3 Analysis of Alternatives contained no analysis of benefits. In fact, it did not identify any benefits of investment beyond the assumption—inherent in its focus on increasing the number of detection cycles per day—that earlier detection has the potential to save lives and limit economic loss, a basic and accepted principle for all enhanced surveillance efforts. In selecting the optimal solution, other costs and benefit factors like the examples shown in table 2 could have been helpful.

Table 2: Selected Additional Cost and Benefit Considerations for Evaluating Gen-3 Alternatives

Selected additional cost considerations	Selected benefit considerations
Cost per additional BioWatch jurisdiction	Risk reduction per additional BioWatch jurisdiction
Cost per fraction of U.S. population covered	Risk reduction per fraction of U.S. population covered
Cost per fraction of BioWatch jurisdiction population covered	Risk reduction per fraction of BioWatch jurisdiction population covered
Cost per hour of detection time	Risk reduction per hour of detection time

Source: GAO.

Identifying benefits and conducting a more complete analysis of cost and benefits would help DHS develop the kind of information that would inform tradeoff decisions and the selection of an optimal solution. For example, DHS plans to deploy about four times as many Gen-3 detectors than Gen-2 detectors, and each Gen-3 detector will test the air more frequently, so the Gen-3 deployment plan will increase the depth and range of the coverage to be provided. Specifically, Gen-3 is expected to cover 90 percent of the population in the jurisdictions where it is deployed, as opposed to 65 percent with the currently deployed technology. On the basis of the estimated annual program cost and the percentage of U.S. population covered, we calculated that the estimated Gen-3 deployment will cost about $4 million annually, on average, for each 1 percent of population covered within BioWatch jurisdictions (about $2.7 million more for each 1 percent covered than Gen-2). Moreover, according to Biowatch program officials, this kind of calculation may actually underestimate the cost of increasing population coverage by deploying more detectors, because the relationship between number of detectors and population covered is not linear. However, because the Gen-3 Analysis of Alternatives does not include a discussion of benefits, or a cost-benefit analysis, it does not consider the extent to which expanding the population covered under the proposed Gen-3 deployment would contribute to a reduction of risk and at what cost.

In its evaluation of BioWatch and public health surveillance, the National Academies stated that the BioWatch program should not expand its coverage of biological agents or jurisdictions without a clear understanding of the change's contribution to reducing mortality or morbidity in conjunction with clinical case finding and public health.[25] However, without a discussion of benefits, the Analysis of Alternatives could not help DHS develop this understanding before approving the acquisition at ADE-2A. In June 2011, the program office commissioned a study from Sandia National Laboratories that began to develop a basis for this kind of understanding of benefits related to public health outcomes.[26] However, critical information related to both the costs and the benefits of the planned Gen-3 approach remains to be explored. DHS has commissioned the Homeland Security Institute (HSI) to conduct an independent study that has as its overarching objective the characterization of the state of Gen-3 technology—that is, whether it is mature enough to continue as an acquisition or whether it needs additional development work.[27] As part of that study, DHS has asked HSI to consider, among other things, (1) whether the threat is adequately described, (2) whether it is possible to determine costs and benefits, and (3) to what extent prior studies like the Sandia study have been validated and used to inform plans for Gen-3 deployment. DHS told us that the study would be completed by September 1, 2012. As of early-September 2012, DHS has not provided us with a copy of the study or responded to requests to provide an updated timeline for the study.

Beyond the uncertainty related to the costs and benefits of the planned Gen-3 approach, there is additional uncertainty about the benefit of this kind of environmental monitoring because as a risk mitigation activity, it has a relatively limited scope. As the study committee for the National Academies evaluation of BioWatch noted, there is considerable

---

[25]See Institute of Medicine and National Research Council, *BioWatch and Public Health Surveillance,* 2011.

[26]*BioWatch Technical Analysis of Biodetection Architecture Performance,* Sandia National Laboratories, January 2012.

[27]The Homeland Security Institute is one of 39 Federally Funded Research and Development Centers. See 6 U.S.C. § 192; 48 C.F.R. § 35.017.

uncertainty about the likelihood and magnitude of a biological attack, and how the risk of a release of an aerosolized pathogen compares with risks from other potential forms of terrorism or from natural diseases.[28] The report also notes that while the BioWatch program is designed to detect certain biological agents (currently five agents) that could be intentionally released in aerosolized form, detecting a bioterrorism event involving other pathogens or routes of exposure requires other approaches. Given the higher total estimated operating cost for the Gen-3 program, it is important, especially in an increasingly resource-constrained environment, to also consider the benefit—in terms of its ability to mitigate the consequences of a potentially catastrophic biological attack—that the extra investment provides. These scope limitations provide context in both the consideration of mission need and in analyzing cost effectiveness.

Because the Gen-3 Analysis of Alternatives focuses on justifying total replacement of Gen-2 technology with an autonomous detection technology, it did not explore whether another solution might be more effective. For example, according to BioWatch program officials, it is possible that other options—including but not limited to deployment of some combination of both technologies, based on risk and logistical considerations—may be more cost-effective. Along these lines, program officials told us that in 2011, to help them manage various budget contingencies, they prepared a summary of available deployment options for Gen-3 that includes a mixed deployment of Gen-3 and Gen-2 units. However, a more comprehensive solution set was not available to be considered at ADE-2A; nor has such an effort since been undertaken to inform investment and trade-off decisions at the departmental level. Given the uncertainty related to the costs, benefits, and risk mitigation potential of Gen-3, DHS does not have reasonable assurance that the strategy of expanding and completely replacing the existing Gen-2 technology with an autonomous detection technology provides the most cost-effective solution.

---

[28]See Institute of Medicine and National Research Council, *BioWatch and Public Health Surveillance*, 2011. The study committee made these observations, noting that the issues encompassed in them are beyond the scope of its study, but nevertheless fundamental to its assessment and recommendations.

DHS Did Not Develop Complete and Reliable Performance, Schedule, and Cost Information before Approving the Gen-3 Acquisition

In October 2009, DHS approved the Gen-3 acquisition at ADE-2A—based on the information contained in acquisition documents provided by the BioWatch program—authorizing the BioWatch program to proceed with characterization testing of Gen-3 candidate technologies.[29] One critical purpose of the ADE-2A documentation set required by DHS's acquisition guidance is to describe the expected performance, cost, and schedule parameters for an acquisition. We reported in June 2010 that stable parameters for performance, cost, and schedule are among the factors that are important for successfully delivering capabilities within cost and schedule expectations.[30] We also reported in May 2012 that without the development, review, and approval of key acquisition documents, agencies are at risk of having poorly defined requirements that can negatively affect program performance and contribute to increased costs.[31] However, the ADE-2A Acquisition Decision Memorandum stated that significant data necessary for the proper adjudication of an ADE-2A decision were missing. Specifically, it noted that the ADE-2A documentation set did not contain three required documents, including: (1) a Concept of Operations—intended to provide critical information on how an acquisition will function in the operational environment, (2) an Integrated Logistics Support Plan—intended to document how an acquisition will be supported and sustained through its life-cycle, and (3) a Life-Cycle Cost Estimate—intended to provide a credible estimate of the life-cycle cost of the acquisition. Additionally, we found that certain information contained in the ADE-2A documentation set on operational requirements, schedule projections, and cost were not developed using reliable methods as discussed later in this report. For more information on the limitations of Gen-3 acquisition documents and processes at ADE-2A, see appendix I.

As was the case for the Mission Needs Statement and the Analysis of Alternatives, BioWatch program officials stated that they had to prepare

---

[29]From May 2010 through June 2011, the BioWatch program completed a series of characterization tests on a candidate Gen-3 technology. This testing consisted of four independent laboratory tests and one field test in a BioWatch jurisdiction. For more information on the specific testing events conducted, see appendix II.

[30]GAO, *Department of Homeland Security: Assessments of Selected Complex Acquisitions*, GAO 10-588SP (Washington, D.C.: June 30, 2010).

[31]GAO, *Homeland Security: DHS and TSA Face Challenges Overseeing Acquisition of Screening Technologies*, GAO-12-644T (Washington, D.C.: May 9, 2012).

ADE-2A documentation quickly because ADE-2A had been accelerated by 14 months. However, in the absence of complete and reliable information, DHS had limited assurance that the acquisition would successfully deliver the intended capability within cost and on schedule. Nevertheless, the Deputy Secretary approved the acquisition, but she required the program office to provide quarterly progress updates.

On the basis of the Gen-3 documentation submitted at ADE-2A, DHS expected to acquire a system that would cost $2.1 billion, be fully deployed by fiscal year 2016, and meet certain performance requirements. As shown in table 3, as of July 2012, the performance, schedule, and cost parameters for the Gen-3 acquisition are significantly different from the parameters DHS expected when it approved the acquisition at ADE-2A.

Table 3: Significant Changes to Performance, Schedule, and Cost Expectations for Gen-3 Acquisition

Acquisition expectations	Expectation at ADE-2A	Current status
Performance	Satisfy five Key Performance Parameters (KPP)—the most important and non-negotiable requirements that a system must meet in order to fulfill its intended purpose.[a]	**One KPP not met; revisions pending.** In response to challenges revealed during Phase 1 characterization testing, the BioWatch program has submitted significant revisions to one KPP—System Sensitivity—to DHS for approval.[b]
Deployment schedule	Initial deployment in fiscal year 2013; full deployment no later than fiscal year 2016	**Delayed.** Current projections estimate that initial deployment will begin in fiscal year 2016, with full deployment in fiscal year 2022.
Life-cycle cost estimate	$2.1 billion	**Increased.** The current estimated life-cycle cost is $5.8 billion.[c]

Source: GAO analysis of BioWatch key acquisition documents.

[a]The five KPPs for the Gen-3 acquisition at the time of ADE-2A included Biological Agent (the number of agents), System Sensitivity (the amount of agent that has to be present for the system to detect it), Time to Detect, Achieved Availability, and Probability of False Positive (the probability that the detector will issue a positive signal when no agent is present).

[b]During characterization testing, the candidate technology tested met the Time to Detect and Achieved Availability KPPs, partially met the Biological Agent KPP, and did not meet the System Sensitivity KPP. Performance against the final KPP—Probability of False Positive—remains unresolved. The BioWatch program has proposed revisions to the KPPs for the next phase of testing, including adding a KPP requiring the system to be autonomous and integrated while removing probability of false positive from the KPPs. The most significant proposed revision was a decrease in the System Sensitivity KPP. For additional information on the KPPs, see appendix II.

[c]The $2.1 billion life-cycle cost estimate (a point estimate not adjusted for risk) submitted in October 2009 for ADE-2A was the estimate used for planning purposes at the time. In the June 2011 Life-Cycle Cost Estimate, the BioWatch program recommended the 80 percent confidence level for planning purposes. In this table, we compare the two estimates used for planning purposes. The point estimate at the 28 percent confidence level in the June 2011 Life-Cycle Cost Estimate is $3.8 billion.

The confidence interval represents the probability that the actual cost will be at or lower than the estimate.

Regarding performance expectations, the BioWatch program submitted a revised Operational Requirements Document to DHS for approval that includes a proposed revision to the key performance parameter for system sensitivity— the amount of a pathogen that would have to be present in the air for the system to detect its presence. DHS acquisitions guidance requires components to develop key performance requirements that an acquisition must meet in order to fulfill the program's fundamental purpose and close the capability gap(s) identified in the Mission Needs Statement, and to document these requirements in an Operational Requirements Document. However, BioWatch program officials told us that the original sensitivity requirement was based on what DHS thought the technology could theoretically achieve, and was not informed by a scientific and risk-informed assessment of what level of sensitivity would be needed—from an operational perspective—to fulfill the Gen-3 purpose of mitigating consequences in the event of a biological attack. Additionally, the process used to set the sensitivity requirement did not reflect stakeholder consensus about how to balance mission needs with technological capabilities. Specifically, the BioWatch program did not prepare a Concept of Operations before ADE-2A. According to DHS acquisitions guidance, in developing a Concept of Operations, stakeholders engage in a consensus-building process regarding how to balance technological capabilities with mission needs in order to gain consensus on the use, capabilities, and benefits of a system. Because DHS did not prepare a Concept of Operations before establishing operational requirements, the sensitivity requirement did not reflect broad stakeholder engagement in balancing schedule, cost, and risk realities with achieving a specified mission outcome—for example, a specific level of population protection.

During characterization testing, the candidate technology tested was unable to meet the original sensitivity requirement. According to the September 2011 Operational Assessment, the system sensitivity demonstrated during characterization testing was orders of magnitude lower than the original requirement, meaning that a significantly greater concentration of a pathogen than specified in the requirement would have to be present in the air to trigger detection.[32] According to BioWatch program officials, the original sensitivity requirement was set based on interest in pushing the limits of potential technological achievement rather than in response to a desired public health protection outcome. They said that this led to a requirement that may have been too stringent, resulting in higher costs and schedule delays without demonstrated mission imperative. [33] Because DHS did not ground the sensitivity requirement in Gen-3 program goals, when the candidate technologies were unable to meet the requirement, DHS encountered delays and uncertainty about how to move forward. In response to these concerns, the BioWatch program directed Sandia National Laboratories to evaluate the level of system sensitivity that would be necessary for the Gen-3 program to fulfill its fundamental purpose. The study, which was completed in January 2012, contained findings that, according to BioWatch Program officials, confirm that the sensitivity requirement could be relaxed without significantly affecting the program's public health mission.[34] In response to this study, the BioWatch program submitted an updated Operational Requirements Document with a revised sensitivity requirement to DHS in March 2012 for approval in preparation for ADE-2B, as shown in figure 6.

---

[32]During characterization testing, the candidate autonomous detection system tested did not meet the system sensitivity performance requirement. For additional information on the five KPPs and other system requirements, and results from developmental testing, see appendix II.

[33]The more stringent the sensitivity requirement, the lower the concentration of a pathogen that must be in the air for the system to detect its presence.

[34]Sandia National Laboratories, *BioWatch Technical Analysis of Biodetection Architecture Performance,* (January 2012).

GAO-12-810 BioWatch Gen-3 Acquisition

Figure 6: Timeline of Characterization Testing and Consideration of Sensitivity Requirement

October 2009	May 2010–June 2011	September 2011	January 2012	March 2012	August 2012	September 2012
DHS officially approved the Gen-3 acquisition	Characterization testing of candidate autonomous detection systems[a]	Operational assessment completed summarizing the results of characterization testing	Sandia study on system sensitivity completed	BioWatch program submitted revised sensitivity requirement to DHS for approval	Acquisition Review Board determined that BioWatch could issue a solicitation for performance testing in the next testing phase. The program must return to the Acquisition Review Board for authorization to issue a contract.	Results of HSI study expected; will inform ADE-2B decision.

ADE 2A

ADE 2B

Legend:

ADE - Acquisition Decision Event
HSI - Homeland Security Institute

Source: GAO analysis of BioWatch program documents.

[a]This testing consisted of four independent laboratory tests and one field test in a BioWatch jurisdiction. For more information on the specific testing events conducted during characterization testing, see appendix II.

The need to reevaluate the sensitivity requirement for the Gen-3 acquisition has contributed to delays in the acquisition schedule. For example, in August 2011, the BioWatch program requested to postpone the ADE-2B, scheduled for September 2011, until December 2011 to give the program time to address the testing issues associated with the sensitivity requirement. Given that the Sandia study was not available until January 2012, the program office again requested that ADE-2B be delayed until March or April 2012. As of September 2012, DHS has not approved the revised sensitivity requirement and plans to revisit that decision at the next Acquisition Review Board for ADE-2B. DHS acquisition guidance states that the accurate definition of requirements is imperative if an acquisition is to be completed within schedule constraints and still meet the component and department's mission performance needs. It follows that these schedule delays could have been mitigated if the original sensitivity requirement had been more realistically set using scientific and risk information to ensure that it aligned with the mission need of the program and balanced mission goals with technological feasibility.

In addition to the impact that changing the sensitivity requirement had on the acquisition schedule, the change in schedule expectations since October 2009 can also be explained by DHS not employing reliable schedule estimation methods to produce the schedule estimate that was submitted with the Acquisition Program Baseline in the ADE-2A

GAO-12-810 BioWatch Gen-3 Acquisition

documentation set. Our prior work has found that realistic acquisition program baselines with stable requirements for cost and schedule are among the factors that are important to acquisitions successfully delivering capabilities within cost and schedule constraints.[35] However, BioWatch program officials told us that they set the ADE-2A schedule estimate aggressively because there was pressure to respond quickly to the call to deploy autonomous detection. Additionally, they reported that they did not account for risk in the schedule estimates that were included in the Acquisition Program Baseline for ADE-2A. The BioWatch program office has revised the acquisition schedule since ADE-2A was held in 2009. The most recent update—completed in January 2012—estimated full deployment of the Gen-3 system in fiscal year 2022, 6 years later than anticipated. While the acquisition is currently on track with the January 2012 schedule, the schedule remains subject to uncertainty, in part because of a pending decision about the acquisition strategy.

In addition to changes in the performance requirements and schedule estimates for the Gen-3 acquisition, the cost estimates have also changed since ADE-2A, primarily because the June 2011 Life-Cycle Cost Estimate was calculated using more reliable methods than those used to complete the ADE-2A cost estimate. The BioWatch program did not complete a full Life-Cycle Cost Estimate for ADE-2A as directed by DHS acquisitions guidance. Instead, the program officials submitted a point estimate of $2.1 billion based on their operational experience with an early prototype system. However, this point estimate was not completed in accordance with the *GAO Cost Estimating Guide*, which DHS uses for cost estimating to help ensure the reliability of its cost estimates.[36] For example, the cost estimate did not account for risk and uncertainty, and it was not based on the work breakdown structure for Gen-3 and as such, DHS did not have assurance that it captured all relevant costs.[37] BioWatch program officials told us that, as was the case with the Mission Needs Statement and the Analysis of Alternatives, they did not have time

---

[35]GAO-10-588SP; GAO-12-644T.

[36]GAO-09-3SP.

[37]A work breakdown structure defines in detail the work necessary to accomplish a program's objectives. It is a necessary program management tool because it provides a basic framework for a variety of related activities including estimating costs, developing schedules, identifying resources, determining where risks may occur, and providing the means for measuring program status.

to engage in a full effort to develop a Life-Cycle Cost Estimate in accordance with the *GAO Cost Estimating Guide* ahead of ADE-2A, but were directed by the department to proceed with the best point estimate they could derive. Additionally, both BioWatch program and PARM officials described a climate before ADE-2A in which the department's business processes—including acquisition practices—were maturing and thus were less rigorous in their adherence to best practices for cost and schedule estimating.

The BioWatch program has revised the cost estimate using more reliable methods since the ADE-2A estimate was prepared in 2009. The most recent update—completed in June 2011—shows the estimated life-cycle cost for the Gen-3 acquisition to be $5.8 billion (80 percent confidence), much higher than the $2.1 billion point estimate presented at ADE-2A. The 2011 Life-Cycle Cost Estimate was aligned with GAO's *Cost Estimating Guide*, which recommends that agencies calculate a range of possible cost estimates based on different risk levels in order to account for uncertainty. According to the guide, experts agree that program cost estimates should be budgeted to at least the 50 percent confidence level, but budgeting to a higher level (for example, 70 percent to 80 percent) is now a common practice. Moreover, a higher confidence level in cost estimating may be more prudent, as experts stress that contingency reserves are necessary to cover increased costs resulting from unexpected design complexity, incomplete requirements, technology uncertainty, and other uncertainties that can affect programs, according to the GAO *Cost Estimating Guide*. Acknowledging the benefit of a higher confidence level for cost estimates, the BioWatch program recommended that the 80 percent confidence level estimate be used for planning purposes. As such, the $5.8 billion figure presented in the 2011 cost estimate was calculated at the 80 percent confidence level—meaning that there is an 80 percent chance that the actual life-cycle cost will be this amount or less, according to BioWatch officials.

BioWatch program officials told us that the large difference between the ADE-2A cost estimate and the June 2011 cost estimate is primarily driven by the inclusion of risk in the June 2011 estimate, rather than by changes to the program. However, these officials also noted other factors that contributed to the difference. For example, the 2009 estimate was not as robust as the 2011 estimate because it was not based on the work breakdown structure for the program. Additionally, because of changes in the schedule estimates, the June 2011 estimate considers costs through fiscal year 2028, whereas the 2009 estimate considered costs through fiscal year 2020. These changes in performance, schedule, and cost, along with maturation in the department's acquisition management process, create an opportunity for DHS to reevaluate the mission need and alternatives in a more comprehensive and systematic fashion, and in accordance with DHS acquisitions guidance, to help ensure that it invests its limited resources in the most cost-effective solution possible. In addition, using comprehensive and systematically developed information, in conjunction with good practices for cost and schedule estimating like those described in the *GAO Cost Estimating Guide*, could help ensure that the department and policymakers have the most reliable performance, schedule, and cost information available for decision making.

BioWatch Must Demonstrate System Performance and Receive Approval before Full Deployment, Estimated for 2022

According to DHS officials the remaining steps in the Gen-3 acquisition include performance testing, operational testing and evaluation, production, deployment, and sustainment. Figure 7 shows the timeline, based on the January 2012 Acquisition Program Baseline and discussions with BioWatch program officials, for the remaining steps to deploy and operate Gen-3.

Figure 7: Estimated Schedule for Key Remaining Gen-3 Deployment Steps (as of September 2012)

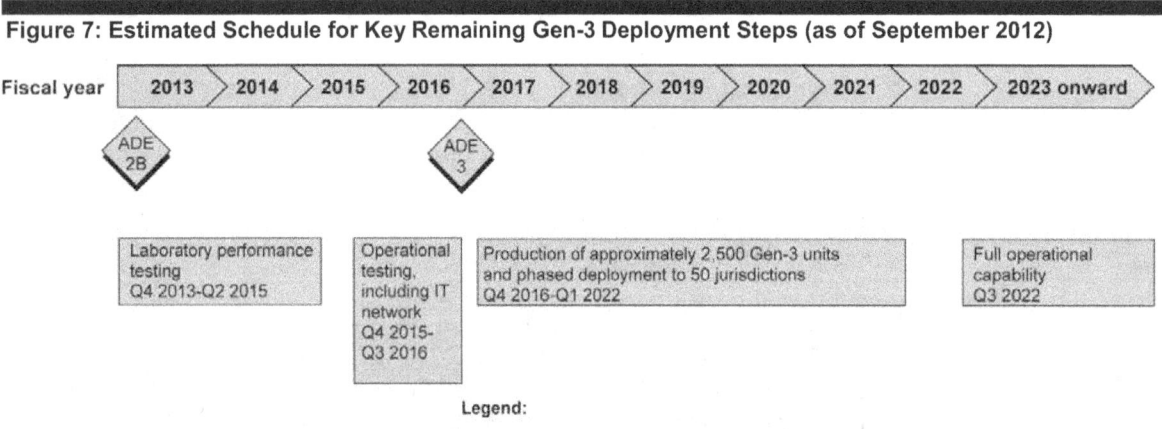

Legend:

ADE - Acquisition Decision Event
Q - Fiscal year quarter

Source: GAO analysis of BioWatch program documents.

First, DHS plans to issue a solicitation for performance testing in the next testing phase, but the Acquisition Review Board must provide approval before the program awards a contract. In addition, final ADE-2B approval will be required for the remaining acquisition steps, including operational testing and evaluation.[38] In preparation for ADE-2B, the BioWatch program has updated key acquisition documents—including the Life-Cycle Cost Estimate and Acquisition Program Baseline— as required by the Acquisition Decision Authority in a February 2012 memo. In order to inform the ADE-2B decision, these documents must accurately reflect changes to Gen-3 performance requirements and updated cost and schedule estimates for the acquisition and therefore may require further revisions.

The next stage after performance testing is operational testing and evaluation. The goal of operational testing and evaluation for the Gen-3 acquisition is to demonstrate full system performance in the operational environment—jurisdictions in which Gen-3 will be deployed—and to build upon characterization testing conducted in 2010 and 2011, which was intended to assess the state of available technology.[39] Characterization testing was designed to demonstrate candidate technology's performance against requirements set by the BioWatch program and primarily consisted of laboratory testing of individual system components, such as the analytical subsystem—the component that tests for pathogens. Characterization testing did not demonstrate the full system or subsystems' ability to detect the five pathogens in an operational environment because of legal restrictions on testing biological pathogens.[40] Additionally, characterization testing did not include laboratory or field testing of the information technology network that will transmit results for public health officials, which has not yet been developed. Operational Test and Evaluation will be the first opportunity for the BioWatch program to fully test the information technology network that will disseminate real-time results to public health officials, a key component of the Gen-3 system. The September 2011 Operational Assessment summarizing the results of characterization testing concluded that without complete testing of the information technology network, aspects of real-time system performance in an operational

---

[38]The next phase of testing is to include performance testing in three independent laboratories and operational test and evaluation in four BioWatch jurisdictions.

[39]For more information on characterization test events and results, see appendix II.

[40]DHS officials told us that legal restrictions prevent them from testing biological pathogens in an operational environment. See, e.g., 18 U.S.C. ch. 10; 7 C.F.R. pt. 331; 9 C.F.R. pt. 121; 42 C.F.R. pt. 73.

environment remain unverified, and failure to demonstrate this capability may seriously inhibit user confidence in the system.

Results from operational test and evaluation will be used to inform ADE-3, which, if approved would authorize full-rate production and deployment of Gen-3. DHS's Acquisition Life-cycle Framework requires that the BioWatch program provide proof that the technology satisfies the operational requirements. To ensure that the full system satisfies the operational requirements, the BioWatch program intends to design a testing plan that demonstrates that the full system—including the information technology network when it is developed—can operate as intended, while complying with legal restrictions on testing for pathogens in BioWatch jurisdictions. DHS has not yet finalized a testing strategy, and the final test plan will depend on the candidate technologies chosen for testing following ADE-2B. Whatever the strategy, DHS officials from the BioWatch program and the Science and Technology Directorate office that oversees testing said that operational test and evaluation will include a number of subsystem and full system test events from which performance in an operational environment can be modeled and extrapolated. Table 4 provides examples of possible test events to demonstrate Gen-3 performance.

Table 4: Possible Test Events Planned to Demonstrate Gen-3's Operational Performance

Test	Description	Estimated date (fiscal year)
Laboratory testing of subsystems at three independent laboratories[a]	Includes evaluating Gen-3 assays, or tests, to measure sensitivity and specificity of the part of the device that tests for pathogens, testing of the aerosol collection subsystem, and testing of the analytical subsystem to determine the lowest concentration of a BioWatch pathogen that the subsystem can detect, including in the presence of contaminants, such as dust or pollen[b]	2013-2015
Laboratory testing of the full system[a]	Uses aerosolized killed agents to confirm the projected system sensitivity.	2013-2015
Environmental testing	Verifies the Gen-3 detector's physical integrity and is to consist of simulating conditions representative of BioWatch jurisdictions, such as temperature and humidity, by operating the detector in an environmental chamber.	2013-2015
Field tests in four BioWatch jurisdictions	Deploys 20 Gen-3 detectors in each jurisdiction expected to be colocated with Gen-2 detectors at a variety of sites representing a cross section of operational environments.	2015-2016

Test	Description	Estimated date (fiscal year)
Information assurance evaluation for information technology network	Evaluates individual components of the information network and the functionality of the full system during field testing. Focus on the confidentiality, integrity, and availability of the network and is to be based on the requirements developed by the National Institute of Standards and Technology.[c]	2015-2016
Logistics demonstration	Demonstrates the readiness of the contractor's System Support Package. The event is to include both remove/replace and diagnostics/prognostics elements of field-level maintenance. The data generated by this event are also to support assessments of the potential cost, schedule, and performance impacts of the systems' suitability risks.	2015- 2016

Source: GAO analysis of BioWatch program documentation.

[a]The extent of laboratory subsystem and full system testing required will depend on whether the candidate technology has been shown to meet some or all of the data requirements in the Requests for proposals.

[b]Sensitivity refers to the amount of agent that must be present for the detector to recognize it. Specificity refers to how well the test distinguishes between the genetic material of the agent it is designed to detect and another—possibly genetically similar—agent.

[c]Department of Commerce, National Institute of Standards and Technology, Federal Information Processing Standards Publication 1999; Standards for Security Categorization of Federal Information and Information Systems (Gaithersburg, Maryland: 2004).

Collectively, the BioWatch program estimates that this testing will take approximately 3 years and cost approximately $89 million. During operational testing and evaluation, the BioWatch program must prepare for and mitigate several limitations. These limitations include the following:

- **Inability to fully test Gen-3's detection capability:** BioWatch officials told us that legal restrictions on the aerosolized release of all five BioWatch agents in U.S. cities limit the BioWatch program's ability to demonstrate full and subsystem performance in an operational environment.[41] Without releasing the agents in BioWatch jurisdictions, the BioWatch program is unable to test the system's ability to detect them in the operational environment. According to BioWatch program officials and DHS S&T officials who assist with test design, designing laboratory and field tests that can compensate for these limitations on pathogen use is a goal that is guiding the development of the testing plan for Operational Testing and Evaluation.

---

[41]See, e.g., 18 U.S.C. ch. 10; 7 C.F.R. pt. 331; 9 C.F.R. pt. 121; 42 C.F.R. pt. 73.

GAO-12-810 BioWatch Gen-3 Acquisition

- **Inconsistent performance in different operational environments:** The candidate system tested during the characterization field test performed better at some sites than others. Specifically, detectors located on underground subway platforms had higher incidences of malfunction than detectors in other locations. These malfunctions may be associated with the presence of metallic brake dust; however, this failure demonstrates different operational environments pose different challenges, and the BioWatch program plans to conduct laboratory testing as well as modeling to further assess detector performance under different operational conditions.

- **Difficulty verifying false positive rate:** In order to build user confidence in the system, the BioWatch program has established a stringent threshold of 1 in 10 million for the false positive rate—that is, the rate at which the system is allowed to indicate a pathogen is present when one is not. However, according to BioWatch documentation, 33.5 years of operational testing would be required to fully demonstrate that the system meets the established false positive rate. Therefore, the BioWatch program plans to use data from laboratory testing to model and extrapolate the probability of a false positive. According to program documentation, the amount of time planned for operational testing will be sufficient to reveal any issues with false positive performance of the candidate technologies tested.

The goal of the next phase of testing is to demonstrate that Gen-3 candidate technologies can operate as intended in the operational environment. To achieve that goal, which is required for ADE-3, the BioWatch program must successfully mitigate these testing limitations. For example, to address the inconsistent performance in testing environments, the program must determine whether and how to adjust laboratory conditions to better reflect the operational environment by exposing the detectors to contaminants such as dust and pollen.

To be ready to produce and deploy Gen-3, DHS must demonstrate technological readiness for the full system based on both individual component readiness and the maturity of the integration of those components. In August 2011, on the basis of results of characterization testing, the Institute for Defense Analysis conducted a Technology Readiness Assessment—a formal independent review that assesses the maturity of critical hardware and software technologies to be used in systems—for Gen-3. Using the Department of Defense's (DOD) Technology Readiness Level (TRL) scale, which defines levels of technological maturity on a scale of 1 to 9, the assessment assigned TRL scores to the Gen-3 candidate technology's individual critical technology elements, that is, those subsystems that are vital to the functioning of the system and are either new or novel applications or pose major technical risk. This assessment rated all but one of the critical technology elements it assessed as TRL 7—indicating a relatively high level of maturity for each technology element assessed. However, the assessment does not provide an overall TRL for the full Gen-3 system. It notes that doing so could obscure the strengths and weaknesses of individual system components, and says that the DOD's *Technology Readiness Assessment Deskbook*, which provides guidance for assigning TRLs, does not describe how to aggregate TRLs. However, other DOD guidance specific to chemical and biological defense says that a TRL evaluation is generally undertaken to establish a system's level of maturity relative to a specific purpose, which suggests that the next phase of testing should result in a technology readiness assessment that provides an indication of how well these components perform together in order to meet the mission need of autonomous detection.[42] Furthermore, we have previously reported that underestimating the complexity of systems integration can be a cause of significant cost and schedule growth.[43] DHS also has not assessed the technology readiness of the data network, a major component of the Gen-3 system, or its integration into the system because it has not yet been developed. The data network and its integration will therefore require demonstration prior to production and deployment of Gen-3.

If the BioWatch program can demonstrate that the candidate technology meets requirements and DHS approves the Gen-3 acquisition at ADE-3, the DHS June 2011 life-cycle cost estimate indicates that Gen-3 is expected to cost $5.8 billion (80 percent confidence) through June 2028. As shown in figure 8, approximately $5.7 billion of this total has not yet been spent and is expected to primarily fund operations once the system is deployed.

---

[42]DOD, *Chemical Biological Defense Program Technology Transition Handbook.*

[43]GAO, *Defense Acquisitions: Despite Restructuring, SBIRS High Program Remains at Risk of Cost and Schedule Overruns,* GAO-04-48 (Washington, D.C: Oct. 31, 2003).

Figure 8: Previous Spending on Gen-3 through Fiscal Year 2011 and Estimated Costs Remaining

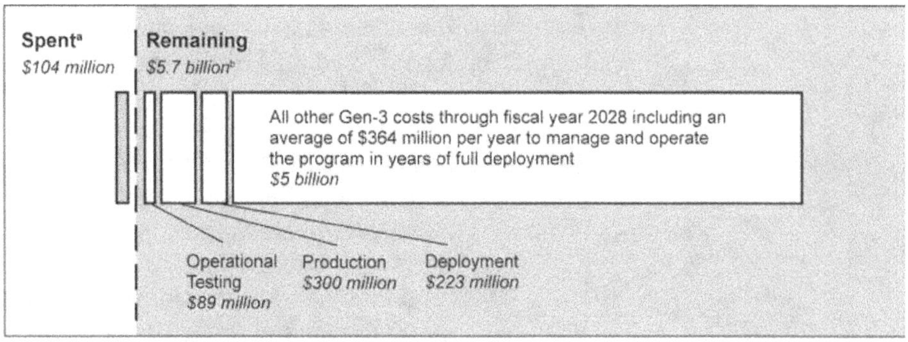

Source: GAO analysis of BioWatch program documents.

[a]Spent cost includes actual costs through fiscal year 2010, as well as estimated costs for fiscal year 2011, as descr bed in the June 2011 BioWatch Gen-3 Life-Cycle Cost Estimate. Spent costs do not include costs for DHS S&T's BAND program.

[b]Remaining costs do not add to $5.7 billion because of rounding.

To prepare for the deployment of Gen-3, the BioWatch program must work with Gen-3 jurisdictions to prepare sites for detector placement and to develop location-specific Concepts of Operations to provide key information and considerations—such as specifying roles and responsibilities and developing public information and risk communication messages—that are integral to response operations in the event that Gen-3 detects a pathogen. Like the Gen-2 system, the Gen-3 system is to be operated by BioWatch jurisdictions, and therefore the system's usefulness in improving response time is expected to be determined, in part, by each jurisdiction's willingness to respond to a positive test result, which, if incorrect could have large monetary costs and public and political repercussions. According to BioWatch program officials, they want the jurisdictions to have enough confidence in the system that they are willing to take action based on positive results from a Gen-3 detector without confirmatory laboratory testing. Therefore, according to BioWatch program officials, they have taken steps to increase jurisdictions' confidence in the Gen-3 system. For example, they provide guidance to jurisdictions and are in the process of developing a quality assurance process to track system performance. Furthermore, these officials anticipate running Gen-2 and Gen-3 concurrently for up to 6 months in BioWatch jurisdictions, and requiring all candidate technologies to archive positive samples so that the jurisdictions can run confirmatory laboratory analysis on the samples.

Conclusions

Despite Gen-3's potential to save lives under specific conditions, uncertainty remains about its general risk mitigation value. DHS established the strategy to quadruple the number of deployed detectors and replace all Gen-2 technology with an autonomous solution while expanding to 20 additional cities without engaging in a robust mission needs effort to serve as a foundation for subsequent acquisition efforts. As we have previously reported, because DHS does not have unlimited resources and cannot protect the nation from every conceivable threat, it must make risk-informed decisions regarding its homeland security approaches and strategies. In addition, we have previously reported that programs that conduct a limited assessment of alternatives before the start of system development tend to experience poorer outcomes than programs that conduct more robust analyses. Without a justified mission need to ground acquisition decision making or a systematic analysis of the cost-benefits and risk, DHS has pursued goals (such as the time threshold of 6 hours) and specific technological requirements (such as the sensitivity threshold) that may or may not support optimal solutions. Reevaluating the mission need and systematically analyzing alternatives based on cost-benefit and risk information could help DHS gain assurance that it is pursuing an optimal solution.

Furthermore, difficulty attaining the original goals has contributed to challenges in meeting milestones and deadlines for deployment. In 2009, when the Acquisition Decision Authority approved the Gen-3 acquisition, it was anticipated that Gen-3 technologies would be in initial deployment by 2013 and fully deployed by the first quarter of 2016. In 2011, DHS's most recent estimate, which contains significant uncertainty because of testing limitations, among other reasons, was that full deployment would be 6 years later, in 2022. Similarly, the $2.1 billion cost estimate presented to DHS decision makers and Congress for planning purposes at the start of the acquisition is now $5.8 billion (for the first 13 years of deployment; only 6 of which are for full deployment) and may still rise because of lingering uncertainty about the acquisition strategy. These changes in cost, schedule, and performance, along with maturation in the department's business processes—including acquisitions and risk management—reinforce the importance and provide an opportunity for DHS to reevaluate the mission need and alternatives in a more robust, considered, and systemic fashion, as called for in the Acquisition Life-cycle Framework, to help ensure that it makes the most sound investments possible. In addition, comprehensive and systematically developed information, developed using good practices for cost and schedule estimating like those described in the *GAO Cost Estimating Guide*, could help ensure that the department and policymakers have the most reliable performance, schedule, and cost information available for decision making.

Recommendations for Executive Action

To help ensure that Gen-3's public health and risk mitigation benefits justify the costs, the program pursues an optimal solution, and DHS bases its acquisition decisions on reliable performance, cost, and schedule information developed in accordance with guidance and good practices, we recommend that before continuing the Gen-3 acquisition, the Secretary of Homeland Security ensure that program and acquisition decision makers take the following two actions:

1. reevaluate the mission need and systematically analyze alternatives based on cost-benefit and risk information, using information from studies like those conducted by the Homeland Security Institute and Sandia National Laboratories, along with any other risk and cost information that may need to be developed, and

2. update other acquisition documents, such as the Acquisition Program Baseline and the Operational Requirements Document, to reflect any changes to performance, cost, and schedule information that result from the reevaluation of mission needs and alternatives.

Agency Comments and Our Evaluation

We provided a draft of this report to DHS for comment, and DHS provided written comments on the draft report, which are reproduced in full in appendix III. DHS also provided technical comments, which we incorporated as appropriate. DHS concurred with both recommendations, but did not concur that these actions need to be completed before continuing with the acquisition.

With respect to the first recommendation to reevaluate the mission need and alternatives, DHS agreed that further evaluation of the mission need and alternatives is necessary. DHS stated that, on August 16, 2012, it directed the BioWatch program to complete an updated Analysis of Alternatives and Concept of Operations, which, according to a DHS official must be completed before ADE-2B, but DHS did not specify how it plans to reevaluate the mission need. With respect to the second recommendation to update other acquisition documents to reflect any performance, cost, and schedule information that might result from reevaluation, DHS acknowledged that it may be necessary and appropriate to do so. However, DHS did not agree that it should implement these two recommendations before continuing the acquisition.

In its response, DHS stated its intent to issue a solicitation for performance testing concurrent with the efforts to implement the recommendations. DHS stated that BioWatch will be required to return to the Acquisition Review Board prior to issuing a contract stemming from this solicitation.

We are pleased that DHS plans to reevaluate the mission need and alternatives and that the department believes this action would be beneficial as it seeks to reduce programmatic risk and demonstrate sound fiscal stewardship in an increasingly constrained fiscal environment. Additionally, we commend DHS's stated commitment to use Acquisition Management Directive 102-01 to ensure consistent and efficient acquisition management, support, review, and approval. The directive's acquisition life-cycle framework is designed to establish a foundation based on critical examination of the capability gap an acquisition would fill and to build sequentially on that foundation to support solid, knowledge-based acquisition decision making. To satisfy the larger purpose of the framework—providing assurance that DHS makes judicious decisions about how to invest limited resources and implements them effectively—it is vital that it be used consistently, that each acquisition adheres to the framework throughout its entire life-cycle, and that specified steps are completed in a sequential manner to support key acquisition decisions.

Accordingly, we are concerned by DHS's intention to continue the acquisition efforts before ensuring that it has fully developed the critical knowledge a comprehensive acquisition life-cycle framework effort is designed to provide. Our work showed that DHS does not have reasonable assurance that the solution it has been pursuing warrants investment of limited resources and that it represents an optimal solution. We believe it is possible that an earnest effort to reconsider the Gen-3 mission needs and alternatives would result in a different plan and course of action than the current effort.

DHS stated in its response that it has directed the BioWatch program to complete an updated Analysis of Alternatives, but it remained silent on what actions, if any, it will take to reevaluate the mission need. As such, it is not clear from DHS's response to what extent it intends to engage in a fresh reevaluation of the mission need in the broader context of DHS's biodefense and related responsibilities before it undertakes efforts to update its Analysis of Alternatives. During discussions with program officials about the recommendation to reevaluate the mission need, the officials told us that they had resubmitted the original Mission Needs Statement to DHS for review. If DHS were to approve the original Mission

Needs Statement and use it to guide the reevaluation of alternatives, it would overlook the intent of the recommendation. The intent is that DHS reevaluate existing capability gaps through the mission needs process to provide a foundation for future acquisition decision making—including the Analysis of Alternatives—that is grounded in better understanding and consensus about how filling these gaps will contribute to larger biodefense needs. Moreover, DHS's plans to pursue testing of the Gen-3 solution—a solution which has driven DHS's efforts for a number of years, including prior efforts to define mission need and analyze alternatives— even while agreeing to reconsider whether it is an appropriate course of action. This plan raises questions about whether the department plans to systematically and objectively reevaluate the mission need and alternatives for fulfilling that need.

As agreed with your offices, unless you publicly announce the contents of this report earlier, we plan no further distribution until 3 days from the report date. At that time, we will send copies to the Secretary of Homeland Security and interested congressional committees. In addition, the report will be available at no charge on the GAO website at http://www.gao.gov.

If you or your staff have any questions about this report, please contact me at (202) 512-8757 or jenkinswo@gao.gov. Contact points for our Offices of Congressional Relations and Public Affairs may be found on the last page of this report. GAO staff who made major contributions to this report are listed in appendix IV.

William O. Jenkins, Jr.
Director, Homeland Security and Justice Issues

Appendix I: Limitations of Gen-3 Acquisition Documents and Processes at ADE-2A

Department of Homeland Security (DHS) Acquisition Management Directive (AMD) 102-01—intended to ensure consistent and efficient acquisition management, support, review, and approval throughout the department—outlines the overall policy and structure for acquisition management at DHS. Specifically, this directive includes document and process requirements for each of the four phases of the department's Acquisition Life-cycle Framework through which DHS determines whether it is sensible to proceed with a proposed acquisition. DHS formally approved the Gen-3 acquisition at Acquisition Decision Event (ADE) 2A in October 2009 without fully using and completing the processes and analyses in Phases 1 and 2 of the Acquisition Life-cycle Framework: (1) identify a capability need and (2) analyze and select the means to provide that capability. However, as shown in table 5, the documentation set DHS had available to inform the ADE-2A decision was incomplete and certain required information was not developed using the most reliable methods available.

Table 5: Limitations of Gen-3 Acquisition Documents and Processes at ADE-2A

Document name[a]	How documents and processes are intended to inform decision making	Status at ADE-2A	Selected information limitations in Gen-3 documents and processes
Mission Needs Statement	Provides a strategic framework for acquisition planning and capability delivery. As the formal description of the top-level need for a capability, it is to provide sufficient detail for reviewers to understand the capability need, but it should not specify a solution. On the basis of the Mission Needs Statement, decision makers are to render an initial decision on whether to proceed at the first decision gate, ADE-1.	◐	The Gen-3 Mission Needs Statement did not reflect a systematic effort to justify a capability need. Rather, it asserted autonomous detection as the only viable solution. Further, the Mission Needs Statement was not finalized until October 2009, just before ADE-2A, after DHS had already signaled its intent to invest in autonomous detection.
Concept of Operations	Describes how the acquired capability will be used in operations. Intended to bridge the capability gap in the Mission Needs Statement to specific performance parameters in the Operational Requirements Document, the concept of operations process is used to gain consensus among stakeholders on the uses, capabilities, and benefits of a system by collaboratively balancing mission goals against the realities of technology, budget, schedule, and risk.	○	A Concept of Operations was not completed prior to ADE-2A. A collaborative process to develop and fine-tune scientific and risk information in order to gain stakeholder consensus on how to balance mission need with technology, budget, schedule, and risk realities did not occur.

Document name[a]	How documents and processes are intended to inform decision making	Status at ADE-2A	Selected information limitations in Gen-3 documents and processes
Analysis of Alternatives	Identifies the optimal solution to meet the capability need by identifying alternative solutions and systematically comparing the alternatives based on cost-benefit and risk information. DHS and the Office of Management and Budget (OMB) guidance call for the consideration of at least three alternatives, in addition to the status quo. DHS guidance also recommends that the Analysis of Alternatives be prepared by a third party to guard against bias.	◐	The Gen-3 Analysis of Alternatives, which was not independently prepared, compared two alternatives and focused on justifying the autonomous detection solution based on cost per detection cycle rather than on indentifying multiple options and systematically evaluating them based on costs, benefits, and risk information.
Operational Requirements Document	Translates the capability need defined in the Mission Needs Statement into operational requirements that complement the approved CONOPS. Its purpose is to identify a number of performance parameters that need to be met by a program to provide a useful capability. Some of those requirements—Key Performance Parameters (KPP)—are so vital to operational success that leadership could consider cancelling or radically revising the program if they cannot be met.	◐	The Gen-3 Operational Requirements Document established five KPPs, but the requirement for one of these—sensitivity— was not set in accordance with DHS acquisition guidance and has proven difficult to achieve.[b] Specifically, the sensitivity requirement was not based on scientific and risk information about the public health benefit of Gen-3 relative to schedule and cost realities. Establishing unrealistic performance requirements that were not grounded in mission need and science created testing challenges, which resulted in schedule delays.
Acquisition Plan	Identifies the strategy by which the selected option will be obtained and supported.	●	The Acquisition Plan was completed in advance of ADE-2A and identified the strategy by which Gen-3 would be obtained and supported. The activities supported by the Acquisition Plan were generally not in the scope of our audit.
Life-Cycle Cost Estimate	Ensures that a credible and documented estimate of all resources for the development, acquisition, fielding, maintenance and disposal of a capability over the system's life-cycle is known.	◐	The Life-Cycle Cost Estimate was not completed prior to ADE-2A. Instead, the BioWatch Program Office submitted a point estimate of $2.1 billion, but this estimate was not reliable because it did not account for risk and uncertainty and was not based on a work breakdown structure consistent with good practices for cost estimates.[c]
Preliminary Integrated Logistics Support Plan	Establishes how the acquisition will be supported and sustained through its complete life-cycle. As the program's overall supportability and sustainment planning document, it should convey the supportability and sustainment strategy in sufficient detail for decision makers to understand and execute the plan.	○	According to BioWatch program officials, the BioWatch program submitted a draft Integrated Logistics Support Plan for review in advance of ADE-2A. However, the Acquisition Decision Memorandum summarizing ADE-2A reported that the Integrated Logistics Support Plan was missing, and directed the program office to develop a satisfactory plan within 60 days. In the absence of such a plan, DHS decision makers did not have information on the supportability and sustainment plan for the acquisition.

Document name[a]	How documents and processes are intended to inform decision making	Status at ADE-2A	Selected information limitations in Gen-3 documents and processes
Acquisition Program Baseline	Documents the program's critical cost, schedule, and performance parameters, expressed in measurable, quantitative terms that must be met to accomplish program goals. The program's Acquisition Program Baseline approval at ADE-2A establishes the formal program/project baseline for cost, schedule, and performance.	◖	The Gen-3 Acquisition Program Baseline approved at ADE-2A included information on performance and cost, but this information reflected the limitations of the Operational Requirements Document and Life-Cycle Cost Estimate discussed above. Additionally, the Acquisition Program Baseline contained a projected acquisition schedule, but this schedule projection was set aggressively, without accounting for risk, as suggested by good practices, including the GAO Cost Estimating Guide.

Legend:

○ Not submitted for ADE-2A.

◖ Submitted for ADE-2A, but contained incomplete or unreliably developed information, or otherwise did not satisfy the purposes established in DHS acquisition guidance.

● Submitted for ADE-2A, and we did not identify deficiencies that relate to the objectives of this report.

Source: GAO analysis of BioWatch Gen-3 acquisition documentation and DHS acquisition guidance.

[a]Other documents discussed in AMD 102-01, including the Capability Development Plan, were not required for the Gen-3 acquisition and have been excluded from this table.

[b]Sensitivity refers to the amount of a pathogen that must be present in the air for the system to detect its presence

[c]A work breakdown structure defines in detail the work necessary to accomplish a program's objectives. According to our Cost Estimating Guide, it is a necessary program management tool because it provides a basic framework for a variety of related activities including estimating costs, developing schedules, identifying resources, determining where risks may occur, and providing the means for measuring program status.

Appendix II: BioWatch Gen-3 Characterization Test Events and Candidate Technology Performance against Key Performance Parameters

From May 2010 to June 2011, the BioWatch program completed a series of characterization tests on a Gen-3 candidate technology.[1] The goals of this testing included characterizing the state of the market and evaluating the candidate systems' abilities to meet performance requirements developed by the BioWatch program. As described in table 6, DHS completed four independent laboratory tests and a field test as part of characterization testing.

Table 6: Test Events Conducted during Gen-3 Characterization Testing

Test event	Description	Location	Date completed
Aerosol collection subsystem test	**Test purpose:** To measure the performance of the candidate technology's aerosol collection subsystem—the system component responsible for collecting air samples. **Results:** The test found that the candidate technology's aerosol collection subsystem had a lower overall sampling efficiency than expected, which negatively affected system sensitivity. The vendor plans to improve sampling efficiency by making an engineering change.	Edgewood Chemical & Biological Center, Maryland	August 2010
Evaluation of assays	**Test purpose:** To measure the performance of Gen-3 assays relative to assays used to test samples from the currently deployed system. Specifically, this test sought to measure the sensitivity—the concentration of a BioWatch agent that can be detected at a specified probability of detection—and specificity—the assay's ability to distinguish BioWatch agents from genetically similar organisms—of the candidate technology's assays in a pristine environment. **Results:** The test found that the sensitivity of the assay was comparable to the reference system with one exception. The assay specificity was comparable to the reference system. However the usefulness of the data is limited to demonstrating the relative performance of the candidate assays against the Gen-2 assays in isolation from all other aspects of system operation such as collection and sample preparation.	Los Alamos National Laboratory, New Mexico	March 2011

---

[1] A second candidate technology participated in two test events—aerosol collection subsystem testing and assay evaluation—but did not complete all testing because the candidate system did not meet program requirements during the assay evaluation Specifically, the second candidate technology yielded both false positives—detecting a BioWatch agent when none was present—and false negatives—not detecting an agent when one was present.

Test event	Description	Location	Date completed
Analytical Subsystem Test	**Test purpose:** • To measure the sensitivity of the analytical subsystem—the component responsible for sorting, preparing and analyzing samples—using live agent spiked samples in order to estimate the Limit of Detection (LOD)—the lowest quantity of a BioWatch agent that can be detected with a given probability of detection. • To measure sensitivity in the presence of environmental contaminants, such as dust and pollen. **Results:** • The analytical subsystem did not perform as expected. Specifically, the LOD was significantly higher than that expected by the vendor, meaning the system required more organisms of a BioWatch agent to trigger a positive result than expected. • The presence of contaminants did not inhibit the candidate technology's ability to detect BioWatch agents or increase the probability of a false positive.	Dugway Proving Ground, Utah	April 2011
System Chamber Test	**Test purpose:** To measure the candidate technology's ability to collect and then analyze BioWatch threat agents in order to estimate the full system sensitivity—that is, the system sensitivity when all subsystems operate together. **Results:** For all BioWatch agents tested, the system was less sensitive than expected, meaning that a greater quantity of BioWatch threat agents had to be present in the air to trigger detection.	Dugway Proving Ground, Utah	May 2011
Field Test	**Test purpose:** To demonstrate the performance of the candidate technology's full system in a representative environment. During the test, 12 candidate Gen-3 detectors operated continuously for approximately 14 weeks in various environments (indoor, outdoor, dirty, clean, etc.) and provided a "proof of concept" of Gen-3 technology in an operational environment. **Results:** The independent group conducting and evaluating the field test found that the candidate Gen-3 technology would be considered suitable for operations in the BioWatch environment with two important caveats: • The system required frequent unscheduled maintenance (on average, the system failed once every 4.5 days), which though relatively easy to repair, would be overly burdensome and expensive in the long run. • The information technology infrastructure, a major component of the Gen-3 system, was not fully tested and therefore it was not possible to fully evaluate the adequacy of information generation, collection and dissemination.	Chicago, Illinois (conducted by the National Assessment Group)	June 2011

Source: GAO analysis of characterization test results

Based on the results of the test events described above, an independent assessment was completed to evaluate the performance of the candidate Gen-3 system tested against the requirements developed by the BioWatch program. These requirements were listed in the Gen-3 operational requirements document, approved at ADE-2A in 2009, and included five key performance parameters (KPP)—the most important and non-negotiable requirements that must be met in order for the program to fulfill its purpose. As shown in table 7, the summary report found that of the five KPPs, the candidate system that completed testing met or partially met three, did not meet one, and that performance against the final KPP remained unresolved.

Table 7: Performance of Candidate Gen-3 System during Characterization Testing against Key Performance Parameters

Key performance parameter[a]	Requirement at ADE-2A[b]	Characterization testing results
1. **Biological agent** Number of BioWatch agents detected	5 existing BioWatch agents	**Partially met** During characterization testing, the candidate system detected four of the five existing BioWatch agents. Legal restrictions prevented DHS from testing the fifth agent.
2. **System sensitivity** The amount of the BioWatch agent that must be present in the air in order for the sensor to detect its presence	60 particles per cubic meter	**Not met** The candidate system tested was significantly less sensitive than required.
3. **Time to detect** Time elapsed between intake of BioWatch agent by the detector and reception of results by public health officials	6 hours	**Met** The candidate system tested reported results within 6 hours.
4. **Achieved availability** Minimum acceptable probability that the detector will function correctly when used under normal conditions in ideal support environment	95 percent	**Met** The system demonstrated an achieved availability of 98 percent. However, the system required frequent maintenance.
5. **Probability of false positive** Maximum acceptable probability that the detector will issue a positive signal when no agent is present	1 in 10 million	**Unresolved** No false positives were reported during characterization testing; however, not enough test cycles were run to confirm that the system met the requirement. To measure the probability of false positive with 95 percent confidence, it would take 33.5 years of operation testing and evaluation

Source: GAO analysis of BioWatch program documentation.

[a]KPPs are those against which the candidate technology was assessed during characterization testing. The BioWatch program has revised several of the KPPs for the next phase of testing. Specifically, the program has added a KPP requiring that the system be autonomous and integrated, and removed the probability of false positive from the KPPs (it is still a requirement). The BioWatch program has also made the system sensitivity requirement less stringent, and changed Achieved Availability to Operational Availability—the probability that the system will be operating or ready to operate at any point in time.

[b]Requirements are the threshold requirement, or the minimum standard that the BioWatch program determined that the candidate technology had to meet in order to achieve program goal.

Appendix III: Comments from the Department of Homeland Security

U.S. Department of Homeland Security
Washington, DC 20528

Homeland Security

August 31, 2012

William O. Jenkins, Jr.
Director, Homeland Security and Justice Issues
U.S. Government Accountability Office
441 G Street, NW
Washington, DC 20548

Re: Draft Report GAO-12-810, "BIOSURVEILLANCE: DHS Should Reevaluate Mission Need and Alternatives before Proceeding with BioWatch Generation-3 Acquisition"

Dear Mr. Jenkins:

Thank you for the opportunity to review and comment on this draft report. The U.S. Department of Homeland Security (DHS) appreciates the U.S. Government Accountability Office's (GAO's) work in planning and conducting its review and issuing this report. Within DHS, one of five core missions is preventing terrorism, which includes preventing the unauthorized acquisition, importation, movement, or use of chemical, biological, radiological, and nuclear materials and capabilities within the United States. To support that mission, BioWatch aids in the prevention of biological terrorism through a nationwide bio-surveillance system designed to detect the intentional release of select aerosolized biological agents.

The Department is pleased with GAO's recognition of the importance of Management Directive (MD) 102-01 to assure consistent and efficient acquisition management, support, review, and approval. This Directive serves to ensure that DHS acquisitions respond to a justified mission need; to select the optimal alternative while balancing cost, schedule, and risk realities; and to ensure that reliable performance, schedule, and cost estimates are available to help guide decision making. The Office of Health Affairs continues to implement the decisions made by the acquisition review board (ARB) and senior Department leadership that provide the appropriate oversight to the BioWatch Generation-3 (Gen-3) acquisition.

The draft report contained two recommendations and the Department fully concurs with these, but the Department does not concur that these recommendations must be completed "before continuing the BioWatch Gen-3 acquisition." This is because the Department has implemented a number of enhancements to significantly improve acquisition management and program execution. This model works to improve the probability of successful program capabilities by employing a program performance evaluation process that continually monitors each major DHS program and evaluates its risk level. There is a direct correlation between an increase in the program's risk threshold and the scheduling of an ARB to mitigate risks before the acquisition proceeds.

Recommendation 1: Re-evaluate the mission need and systematically analyze alternatives based on cost-benefit and risk information, using information from studies like those conducted by the Homeland Security Institute and Sandia National Laboratories, along with any other risk and cost information that may need to be developed.

Response: Concur. DHS agrees that further evaluation of bio-surveillance mission need and bio-detection analysis of alternatives is necessary. The Department conducted an ARB on August 16, 2012, after which an Acquisition Decision Memorandum (ADM) was issued directing BioWatch to complete an Analysis of Alternatives (AOA) and complete a Concept of Operations (CONOPS). The ADM also authorized BioWatch to release the Phase II Gen-3 performance testing solicitation, pending completion of specified prerequisites. This addresses the risk factors without creating a substantial delay in obtaining this important bio-defense capability for the Nation.

The Phase II Gen-3 proposals may provide a better understanding of the current state of technology, which could include information valuable for the CONOPS. The ADM also instructed BioWatch Gen-3 to return to the ARB for a decision to approve awarding the Phase II Gen-3 performance testing contract. Further, when an ADM is issued that allows Phase II Gen-3 performance testing to proceed, it will only authorize a contract to acquire a small number of test units and does not establish a procurement of Gen-3 units for operational use.

Recommendation 2: Update other acquisition documents, such as the Acquisition Program Baseline and the Operational Requirements Document, to reflect any changes to performance, cost, and schedule information that result from the reevaluation of mission needs and alternatives.

Response: Concur. DHS acknowledges it is necessary and appropriate to update acquisition documents per MD 102-01, including the Acquisition Program Baseline and the Operational Requirements Document, to reflect any changes to performance, cost, and schedule information that result from the reevaluation of DHS's mission needs and the AOA. As previously noted, the BioWatch program must return to the ARB for a decision to approve awarding the Phase II Gen-3 performance testing contract.

Again, thank you for the opportunity to review and comment on this draft report. Technical comments were previously provided under separate cover. Please feel free to contact me if you have any questions. We look forward to working with you in the future.

Sincerely,

Jim H. Crumpacker
Director
Departmental GAO-OIG Liaison Office

2

Appendix IV: GAO Contact and Staff Acknowledgments

GAO Contact	William O. Jenkins, Jr., (202) 512-8757 or jenkinswo@gao.gov
Staff Acknowledgments	In addition to the contact named above, Edward George, Assistant Director; Kathryn Godfrey; Allyson Goldstein; and Katy Trenholme made significant contributions to the work. Harold Brumm, Nirmal Chaudhary, Michelle Cooper, Marcia Crosse, Katherine Davis, Amanda Gill, Eric Hauswirth, Tracey King, Susanna Kuebler, David Lysy, Amanda Miller, Jan Montgomery, Jessica Orr, Katherine Trimble, and Teresa Tucker also provided support.

GAO's Mission	The Government Accountability Office, the audit, evaluation, and investigative arm of Congress, exists to support Congress in meeting its constitutional responsibilities and to help improve the performance and accountability of the federal government for the American people. GAO examines the use of public funds; evaluates federal programs and policies; and provides analyses, recommendations, and other assistance to help Congress make informed oversight, policy, and funding decisions. GAO's commitment to good government is reflected in its core values of accountability, integrity, and reliability.
Obtaining Copies of GAO Reports and Testimony	The fastest and easiest way to obtain copies of GAO documents at no cost is through GAO's website (www.gao.gov). Each weekday afternoon, GAO posts on its website newly released reports, testimony, and correspondence. To have GAO e-mail you a list of newly posted products, go to www.gao.gov and select "E-mail Updates."
Order by Phone	The price of each GAO publication reflects GAO's actual cost of production and distribution and depends on the number of pages in the publication and whether the publication is printed in color or black and white. Pricing and ordering information is posted on GAO's website, http://www.gao.gov/ordering.htm. Place orders by calling (202) 512-6000, toll free (866) 801-7077, or TDD (202) 512-2537. Orders may be paid for using American Express, Discover Card, MasterCard, Visa, check, or money order. Call for additional information.
Connect with GAO	Connect with GAO on Facebook, Flickr, Twitter, and YouTube. Subscribe to our RSS Feeds or E-mail Updates. Listen to our Podcasts. Visit GAO on the web at www.gao.gov.
To Report Fraud, Waste, and Abuse in Federal Programs	Contact: Website: www.gao.gov/fraudnet/fraudnet.htm E-mail: fraudnet@gao.gov Automated answering system: (800) 424-5454 or (202) 512-7470
Congressional Relations	Katherine Siggerud, Managing Director, siggerudk@gao.gov, (202) 512-4400, U.S. Government Accountability Office, 441 G Street NW, Room 7125, Washington, DC 20548
Public Affairs	Chuck Young, Managing Director, youngc1@gao.gov, (202) 512-4800 U.S. Government Accountability Office, 441 G Street NW, Room 7149 Washington, DC 20548

Please Print on Recycled Paper.